A CRUDE LOOK
AT THE
WHOLE

ALSO BY JOHN H. MILLER

Complex Adaptive Systems

A CRUDE LOOK
AT THE
WHOLE

The Science of Complex Systems
in Business, Life, and Society

John H. Miller

BASIC BOOKS
A Member of the Perseus Books Group
NEW YORK

Published by Basic Books,
A Member of the Perseus Books Group

Books published by Basic Books are available at special discounts
for bulk purchases in the United States by corporations, institutions,
and other organizations. For more information, please contact the
Special Markets Department at the Perseus Books Group, 2300 Chestnut
Street, Suite 200, Philadelphia, PA 19103, or call (800) 810-4145,
ext. 5000, or e-mail special.markets@perseusbooks.com.

Designed by Linda Mark

Library of Congress Cataloging-in-Publication Data

Miller, John H. (John Howard), 1959–
A crude look at the whole : the science of complex systems
in business, life, and society / John H. Miller.
pages cm
Includes bibliographical references and index.
ISBN 978-0-465-05569-2 (hardcover)—
ISBN 978-0-465-07386-3 (e-book)
1. System analysis. 2. System design. 3. Information modeling.
4. Coupled problems (Complex systems) I. Title.
QA402.M495 2015
003—dc23
2015027692

10 9 8 7 6 5 4 3 2 1

To my parents

Contents

List of Illustrations   *ix*

Preface   *xi*

Prologue   *xvii*

ONE   Introduction:
True Places   1

TWO   From So Simple a Beginning:
Interactions   21

THREE   From Flash Crashes to
Economic Meltdowns:
Feedback   47

FOUR   From One to Many:
Heterogeneity   69

FIVE   From Six Sigma to Novel Cocktails:
Noise   79

SIX   From Scarecrows to Slime Molds:
Molecular Intelligence   99

SEVEN   From Bees to Brains:
         Group Intelligence   113

EIGHT   From Lawn Care to Racial Segregation:
         Networks   141

NINE   From Heartbeats to City Size:
         Scaling   157

TEN   From Water Temples to Evolving Machines:
         Cooperation   169

ELEVEN   From Stones to Sand:
         Self-Organized Criticality   195

TWELVE   From Neutrons to Life:
         A Complex Trinity   205

         Epilogue: The Learn'd Astronomer   227
         Index   241

List of Illustrations

Figure 2.1	A CAPTCHA Challenge	23
Figure 2.2	A Pattern from Rule 30	25
Figure 2.3	A Sea-Snail Shell Pattern	26
Figure 2.4	A Pattern from Rule 22	29
Figure 2.5	Supply and Demand in a Simple Market	33
Figure 2.6	Prices Arising in Some Experimental Markets	42
Figure 3.1	Major US Market Indices on May 6, 2010	48
Figure 3.2	The Price and Volume of E-mini Contracts on May 6, 2010	58
Figure 4.1	Honeybee Workers Cooling a Hive	72
Figure 5.1	A One-Dimensional Search Problem	83
Figure 5.2	Search Using Simulated Annealing	89
Figure 6.1	Chemotaxis Search in a Simulated Bacterium	102

Figure 6.2   Irrationality Induced by
                   Irrelevant Alternatives   108
Figure 7.1   A Swarm of Honeybees   117
Figure 7.2   Swarm Dances over Time   120
Figure 7.3   Likelihood of a Quorum Forming
                   for the Best Choice   123
Figure 7.4   Risk Aversion in Decentralized
                   Decision Making   125
Figure 7.5   A Circular Mill of Ants   134
Figure 8.1   The Community of Lakeland   142
Figure 8.2   The Dynamics of Lakeland   147
Figure 8.3   A Small World Network   151
Figure 8.4   The Schelling Segregation Model   154
Figure 9.1   Metabolic Scaling   161
Table 9.1   Deaths in Warfare, 1820–1945   165
Figure 10.1 Balinese Rice Terraces   170
Figure 10.2 Offerings to the Goddess
                   of the Lake, Bali   176
Figure 10.3 A Simple Two-State Automaton   184
Figure 11.1 Self-Organized Criticality
                   in a Sand Pile   197
Figure 12.1 A MCMC Frog   217

Preface

The great stillness in these landscapes
that once made me restless seeps into me
day by day, and with it the unreasonable
feeling that I have found what I was
searching for without ever having
discovered what it was.

—PETER MATTHIESSEN, "The Tree
Where Man Was Born"

OVER THE PAST TWO DECADES I'VE HAD THE GREAT PLEASURE OF participating in a wonderful scientific endeavor. Whether this quest is on the fringe or the frontier of science may well depend on where you stand and the direction in which you are looking. When I started down this path, I took to heart Thomas Pynchon's edict that "we have to look for routes of power our teachers never imagined, or were encouraged to avoid," and I embraced a new style of modeling that used the ever-growing power of computation on problems that heretofore had been too complex to analyze. My goal was to focus on the big problems that had motivated me to pursue science in the first place, notwithstanding the constant pressure in graduate school and beyond to redirect such inquiries down a narrow path prescribed by the prevailing paradigm.

In 1988 I was fortunate to join a small group of like-minded thinkers hiding out in the high deserts of New Mexico. From such modest beginnings a new wave of complex-systems thinking emerged. Given the heavy investment of most academic institutions and scientists in traditional paradigms and fields, this new work was easily dismissed at first. This dismissal turned out to be rather fortunate, as it allowed an ever-growing group of creative and talented scientists—each of whom for one reason or another felt the need to think differently—to escape the bounds dictated by the academic establishment and to create new forms of scientific inquiry and institutions better suited to taking on the important problems in the world. Our group formulated problems around core ideas, such as adaptation and robustness, rather than traditional academic fields. We embraced a new set of tools made possible by the information age and developed new methods to move beyond the nineteenth-century toolbox used by most scientists. We created new forms of academic institutions, such as the Santa Fe Institute, that embodied the revolutionary mind-set that was fermenting, allowing the easy interchange of ideas, examples, and tools across formerly isolated academic fields. The act was outrageous enough that the traditional academic powers ignored our activities, giving us the needed time to refine our ideas and methods so that we could start to seriously challenge the prevailing norms.

In the intervening years, the field of complex systems has had time to coalesce. Complex systems has always been a field that transcends the usual academic boundaries. Yet, across this vast array of science, a small set of key ideas has emerged, and it is these ideas that will be the focus of this book. My own interests center around complex social systems—that is, systems composed of interacting, thoughtful (but perhaps not brilliant) agents—and most of the examples presented here will be drawn from this domain.

Since the field of complex systems is rapidly evolving, this book is about both the known and the possible. Thus, some of the work discussed here is well grounded in long-standing research efforts, while other parts are of a more speculative nature. My hope is that the combination will convey the excitement of the ongoing quest while also establishing the future prospects for the complex-systems point of view. Of course, any such excursion will, by necessity, be a selective swath through a large field of existing ideas.

Some of the research discussed in this book is the result of past or ongoing collaborations with Simon DeDeo, Russell Golman, Steve Lansing, Scotte Page, Tom Seeley, Michele Tumminello, and Ralph Zinner. Discussions with Walter Fontana, Van Savage, and Geoffrey West have also been instrumental in refining some of the material. Moreover, the various threads of thought weaving their way throughout this work have benefited from discussions with, and encouragement from, Phil Anderson, Ken Arrow, Brian Arthur, Bob Axelrod, Ted Bergstrom, Ken Boulding, Jim Crutchfield, Robyn Dawes, Doyne Farmer, Paul Fischbeck, Murray Gell-Mann, John Holland, Erica Jen, Stu Kauffman, Steven Klepper, Blake LeBaron, George Loewenstein, Cormac McCarthy, Norman Packard, Richard Palmer, John Rust, Cosma Shalizi, Carl Simon, Herb Simon, Peter Stadler, and Hal Varian. Robert Hanneman, Steve Lansing, Baldomero Olivera, Jacob Peters, Tom Seeley, and Geoff West were all kind enough to provide some of their research materials to generate some of the figures. Laurence Gonzales undertook a careful reading of the manuscript, as did my editor, T. J. Kelleher, and I'm grateful to both of them for their suggestions. During the final stages of the book, Sue Warga and Melissa Veronesi provided key contributions. Finally, thanks to my agent, Jim Levine, for championing this project.

I've also been fortunate to participate in two remarkable scientific institutions: Carnegie Mellon University (CMU) and the Santa Fe Institute (SFI). Both places have the same ethos, namely, to find incredibly creative and smart people and put them in an environment that encourages answering the important questions by collaborating across the usual boundaries while minimizing institutional distractions. Maintaining such an environment is hard, and I'm thankful for farsighted and entrepreneurial administrators, such as SFI's founder, George Cowan, who create such academic playgrounds. After a long stint as a department head at CMU, I've come to recognize the challenges of making such institutions work, and I'm grateful to Jerry Sabloff (president of SFI), Jennifer Dunne and Doug Erwin (current and former, respectively, chairs of faculty of SFI), Mark Kamlet (former provost of CMU), and John Lehoczky (former dean of CMU), who devote a remarkable amount of energy to making such institutions work. Other key folks at SFI who have been helpful include Marcella Austin, Patrisia Brunello, Ronda Butler-Villa, Juniper Lovato, Nate Metheny, Ginger Richardson, Janet Rubenstein, Hilary Skolnik, Laura Ware, and Chris Wood. At CMU I headed the Department of Social and Decision Sciences (aka the department that studies interacting and thoughtful agents), and I'm grateful for the wonderful group of colleagues who have surrounded me while in Pittsburgh. Writing a book and running a department are not always compatible activities, and my business manager, Sarah Bernardini, has been gracious and productive throughout this process; I'm thankful to her, to my assistant, Mary Anne Hunter, and to my other staff members for their help throughout the years.

Finally, thanks to my family and the "Lower-Waldron Commune," my friends and neighbors in Pittsburgh, who allow me to participate in a remarkable and vibrant community that demonstrates daily the right purpose and wonderful promise of complex social systems.

J. H. Miller,
August, 2014, Tesuque, New Mexico

Prologue

Humanity today is like a waking dreamer,
caught between the fantasies of sleep
and the chaos of the real world. The mind
seeks but cannot find the precise place
and hour. We have created a Star Wars
civilization, with Stone Age emotions,
medieval institutions, and godlike
technology. We thrash about. We are
terribly confused by the mere fact of our
existence, and a danger to ourselves and
to the rest of life.

> —E. O. WILSON, *The Social
> Conquest of Earth*

Better a cruel truth than a comfortable
delusion.

> —EDWARD ABBEY

COMPLEXITY ABOUNDS.

Yet our traditional scientific way of thinking relies
on reductionism, an idea that has given us both the Archimedean tools to move the world and the delusion that we

understand what we are doing. We inhabit a world where even the simplest parts can interact in complex ways, and in so doing create an emerging whole that exhibits behavior seemingly disconnected from its humble origins. This is at once both a magical and dangerous world, where out of simple beginnings can emerge either a marvelous outcome or an awe-inspiring catastrophe.

By its very nature, emergent behavior is easy to anticipate but hard to predict. Sometimes emergence coincides with our needs. Markets may create prices that transmit a vast array of critical information, resulting in the allocation of goods and services to their best use. At other times emergence works against us. The same markets may inadvertently start to feed on one another, creating a sequence of crashes and altered expectations that cripple a world economy and impact the lives of billions for years.

So complexity abounds, and the same complex powers that gave us life on earth and the ability to think have also allowed us to create productive systems that, on occasion, go terribly wrong. Maddeningly, even when we try to anticipate such failures and build in mechanisms to keep our systems under control, we necessarily increase the level of complexity in the system and create new paths for failure. Whether we are trying to engineer physical systems, such as nuclear power plants, spaceships, or bridges, or engineer social systems, such as health care, tax policy, or food supplies, we are creating systems that will fail in unanticipated ways.

Indeed, to think that we might create complex systems that do only good is a delusion. That said, complex systems that work well provide so many benefits that we are (and ought to be) willing to accept some occasional failures. A temporary flash crash across a few markets may be a small price to pay for the countless benefits that accrue to society when those same markets work well.

When complexity abounds, there be dragons. Nonetheless, it's better to encounter the dragons you know than the ones you don't. Thus, devoting some of our scientific enterprise to understanding better how complex systems work—and, we hope, in that process learning how they can be created and controlled—is a critical investment as we advance into a world of hyperinteractivity. To survive this looming age of complexity, we need to become proactive rather than reactive. In response to the flash crash of 2010, the Securities and Exchange Commission implemented new "circuit breakers" in the trading of stocks on some markets, yet this policy was driven far more by intuition than by insights from a scientific test bed. In the wake of the 2008 financial meltdown, we implemented various stress tests on banks to try to prevent an individual bank failure, yet it is the systemwide connections that lead to ruin.

Ironically, the same computational and communication advances that are driving the complexity of our era are the same tools that may give us the necessary power to understand, and perhaps even harness, that complexity. Computers provide a new window from which to observe and experiment on complex systems. Moreover, our newfound ability to communicate and collaborate rapidly across previously insurmountable distances may accelerate the needed pace of scientific discovery and innovation.

In the past, the term *complex* was used to describe phenomena that were beyond our understanding and, by implication, beyond our ability to influence. This labeling served as a convenient crutch for scientists (and politicians) to dismiss whole swaths of some of the most critical problems facing society, ranging from climate change to financial collapse to terrorism. However, as discussed in the chapters that follow, complexity is an aspect of nature that *is* amenable to scientific analysis, understanding, and perhaps even control. Once this is recognized, a vast frontier of discovery opens up, allowing us to finally make sense of our world.

We find ourselves in a race for knowledge and control of the complex world around us. This is a race that we must win if we are to thrive, and perhaps even survive, as a species. Our very existence relies on the complex systems that bind our food supplies to our energy networks to our global climate to every institution in our society. We have grown to a sheer size and degree of connectivity where local actions now have global consequences.

We thrash about, with the potential of emergent bliss or disaster with every twitch.

ONE

Introduction: *True Places*

> It is not down in any map; true places
> never are.
>
> —HERMAN MELVILLE, *Moby Dick*

S CIENCE IS ABOUT MAPMAKING. IT'S ABOUT TAKING A COMPLICATED world and reducing it to some sparse set of markings on a map that provides new guidance across an otherwise incomprehensible, and potentially hostile, landscape. A good map eliminates as much spurious information as possible, so that what remains is just enough to guide our way. Moreover, when the map is well made we gain a deeper understanding of the world around us. We begin to recognize that rivers flow in certain directions, towns are not randomly placed, economic and political systems are tied to geography, and so on.

Maps—and science—are often more about what we leave out than what we put in. As Jorge Luis Borges catalogs in his one-paragraph-long short story "On Exactitude in Science," "The Cartographers Guilds struck a Map of the Empire whose size was that of the Empire, and which coincided point for point with it. The following Generations, who were not so fond of the Study of Cartography as their Forebears had been, saw that that vast Map was Useless."

Different maps—even of the same landscape—provide different insights into the world. A topographic map provides information on the various hills and dales in the world in just enough detail to be useful to a hiker. A road map, with its sparse set of major cities and the roads that connect them, provides just enough information for a cross-country drive. Divorcing a map from its purpose inevitably leads to frustration. Too little of the right kind of detail, or too much of the wrong kind, encumbers our ability to understand the world.

Science has proceeded by developing increasingly detailed maps of decreasingly small phenomena. At the heart of this reductionist strategy is a hope that once we have detailed maps of the smallest of parts, we can paste the mosaic together and have a useful map of Borges's Empire. That strategy fails, and while the result might please Borges's Cartographers Guild, the mosaic is as much a fool's errand as Borges envisioned.

The problem lies not in the incompleteness of our knowledge but in the dream—no, the fallacy—of reductionism. Reductionism fails because even if you know everything possible about the individual pieces that compose a system, you know very little about how those pieces interact with one another when they form the system as a whole. Detailed knowledge of a piece of glass does not help you see, and appreciate, the image that emerges from a stained-glass window.

Over the past few decades a new science has been brewing. It is a science that recognizes that there are fundamental principles governing our world—such as emergence and organization—that appear in various guises across all of the nooks and crannies of science. For example, in physics, individual atoms organize into magnets, in biology, cells organize into organisms, and in economics, traders organize into markets. The universality of these principles was a surprise to scientists accustomed to thinking in terms of scientific disciplines, and by necessity, this new science transcends the traditional boundaries imposed by our current academic institutions. It is a science where simple things produce complexity and complex things produce simplicity. It is a science that embraces new investigative tools, such as computers serving as modeling substrates, in order to escape the bounds imposed by our usual collection of scientific tools, such as the various pieces of mathematics, largely derived in the late 1600s, that we so often rely on today. More fundamentally, it is a science that challenges our traditional notion that understanding comes from reducing things to their simplest components.

Alas, the new science we are after, the one that may hold sway over critical aspects of our life and destiny, is, as Herman Melville says, "not down in any map; true places never are." Science as currently practiced—with psychology separate from economics, physics separate from biology, and on and on—has been remarkably productive. The creative destruction of scientific ideas, with its inherent quest to define the frontier by publicly disclosing, evaluating, and correcting ideas, has provided us with an engine of insight. The cost, however, is that individual fields have become increasingly separated from one another intellectually. Taking an exact look at a small piece of the world has become the academic norm and has almost fully displaced taking what my Santa Fe Institute colleague Murray Gell-Mann calls "a crude look at the whole."

That may seem a minor problem, but we see its importance when we look at the true places we wish to explore. Take any global-scale, societal challenge, such as financial collapse, climate change, terrorism, epidemics, revolution, or social change: not one neatly aligns with any particular academic field. Moreover, even if one did, the reductionist approach still may not let us understand the whole. The fundamental principles of complexity describe how even simple parts, once together, seemingly take on a life of their own. Having intimate knowledge of, say, each part of an engine, every bolt, piston, cam, and so on, tells us little about what happens when we put those pieces together and they begin to interact with one another. Moreover, such intimate knowledge gives us no insight into what would happen to the engine as a whole if, say, we increase the size of one of the cylinders.

Reduction gives us little insight into construction. And it is in construction that complexity abounds.

From agoras to amoebas, from bees to brains, from cities to collapse, and on up to zebra stripes, the world around us is an encyclopedia of complexity. Sometimes this complexity arises shaped by natural forces such as evolution, as in the consciousness that emerges from our brains. At other times we have a hand in its creation, as in the steady stream of prices that arises from the seemingly chaotic noise and gestures in a commodities trading pit. Without a science of complex systems, we have little chance to understand, let alone shape, the world around us.

The initial academic discussions of complex systems can be traced back to at least 1776, when Adam Smith, in his *Wealth of Nations*, briefly discusses the "invisible hand" as a force that leads self-interested traders to unintentional, socially desirable outcomes. Of course, scientific propositions that are based on an invisible hand are more akin to the invocation of a deity than to a scientific theory and are about as useful to an economist as one of Rudyard Kipling's just-so stories is to a biologist trying to explain how a leopard gets its spots.

The modern movement of complex-systems thinking can be tied to the beginnings of the atomic and information ages, when scientists such as Stanislaw Ulam and John von Neumann, using some of the world's first programmable electronic computers, began to blur the lines between traditional academic fields as they pursued questions such as whether a machine could be truly self-reproducing. Out of this effort arose a class of models that, starting with a collection of simple, well-defined pieces and interactions, results in a surprisingly rich set of global patterns.

The study of those patterns was an important step toward understanding not just the purpose of an animal's markings—say, camouflage—but also how they arise. Is it necessary that there be some master plan contained within the DNA of a leopard that specifies the color of each location on its skin, similar to how a digital image file directs the color of each pixel on a computer display, or is there a more universal explanation that can tell us how a leopard gets its spots?

The simple mathematical and computational models begun by Ulam and von Neumann have given us a lens through which to look at the origins of such complexity. We find that the combination of simple pieces, locally interacting with one another, is sufficient to lead to global behavior that is rather alien to its origins. Thus, the likely answer to how the leopard gets its spots—or how a lowly (but dangerous) sea snail gets its shell pattern, or even how the cacophony of a trading pit results in a well-organized set of trades and prices—is at once far simpler, far more universal, and far more fascinating than we might imagine.

Over the last few decades, the study of interacting systems has opened up a new frontier in our understanding of complex systems. Whether we consider abstract models running at the speed of light inside a computer or the carefully curated anthropological evidence of a century of rice farming, a small set of core principles governing complex systems has emerged. Interacting systems develop feedback loops among the agents, and these loops drive the system's behavior. Such feedback is moderated or exacerbated depending on the degree of heterogeneity among the agents. Interacting systems also tend to be inherently noisy, and such randomness can have surprising global consequences. Of course, who interacts with whom is a fundamental property of these systems, and such networks of interaction are an essential element of complex systems.

Core principles such as feedback, heterogeneity, noise, and networks can be used to understand new layers of complexity. For example, there are complex systems, such as your mind, that generate coherent and productive decisions in a completely decentralized manner, seemingly without control. Other systems, facing deeply embedded constraints such as getting oxygen to all of the cells in your body, lead to scaling laws that can take seemingly disconnected parts of the world and align them along a simple relationship. Yet other systems, such as the members of a social movement, self-organize into critical states that begin to exhibit a common characteristic behavior. Many interacting systems develop cooperation among the agents, a complex behavior that, once arisen, allows agents to shift into a new realm of opportunity, and we are now in a position to understand such a transition. Finally, by repurposing methods and ideas that were first developed at the dawn of the modern science of complex systems, we can generate a new theorem about the behavior of adaptive systems.

These core principles driving complex systems, and their application to understanding new layers of complexity, are the focus of the pages that follow.

One critical aspect of interactions is feedback. Sometimes feedback stabilizes the system, as happens when we install a not-too-touchy thermostat to control the furnace. Other times feedback causes a system to go out of control, as happens when we place a microphone too close to a loudspeaker, producing an ever-increasing screech. The recent growth of market interconnectivity has resulted in a system rife with feedbacks. This has come about through new communication links, the rise of derivative securities, and the use of high-speed, computer-automated trading. Indeed, these changes have easily outpaced our ability to truly understand their implications, and financial markets have become an unintentional Promethean experiment upon which we now base our economic livelihood.

An example is the "flash crash" of May 6, 2010, when a simple oversight in the programming of a trading computer in a suburb of Kansas City, Kansas, caused a temporary collapse of global markets. The havoc that ensued resulted in dramatic price changes on key stock indices and caused the shares of previously valuable mainstay companies to be sold for pennies (not pennies on the dollar, mind you, but pennies). A five-second trading pause invoked fifteen minutes into the crisis was, fortunately (and remarkably), sufficient to begin to restore the system, and the markets settled back into a more familiar pattern.

In 2008 a much larger event happened that resulted in a financial tsunami that rolled across the world's economy, affecting the lives of billions and plaguing us to this day. Examining this crisis, we find a situation where any single agent on the economic landscape, from homeowners to mortgage brokers to rating agencies, was making sane decisions, yet the connections among these agents created a series of unfortunate feedback loops that destined the system to fail.

The economic collapse of 2008 represents a major failure for the profession of economics. Not only did economists fail to see the onslaught coming, but once the crisis arose, they had no idea how to deal with it. Part of this failure can be traced to the reductionist desire to break things down to simple parts. In the language of modern economic theory, this led to a reliance on "representative agents," constructs that attempt to capture the behavior of, say, all consumers using a single megaconsumer. In part, such a choice arises from the fourteenth-century friar Father William of Ockham's dictate to prefer simpler explanations to more complicated ones. Of course, Ockham still requires that the model, complicated or not, explain what we want to understand. In reality, the use of representative agents is also driven by the limitations of the modeling tools typically used by economists, as these tools can be deployed only if there is a high degree of homogeneity in the system.

While homogeneity is a useful assumption—for both philosophical and practical reasons—the study of complex systems suggests that the behavior of heterogeneous systems may not be so easily averaged out. Whether we are looking at the temperature control of a honeybee hive or the likelihood of a riot, heterogeneous systems often function in ways that are different from homogeneous ones.

Recognizing heterogeneity not only changes our predictions about how a system will behave but also alters our policy prescriptions. Homogeneous systems tend to undergo rapid changes and oscillations, while heterogeneous ones tend to react more slowly. Thus, your ability to start, or quash, a social movement is tied to the degree of heterogeneity among the people involved. Similarly, markets may require some heterogeneity among the traders to remain stable.

COMPLEX SYSTEMS OFTEN HAVE SOME INHERENT DEGREE of randomness tied to the behavior of the agents or the structure of interactions. Perhaps surprisingly, such randomness can be useful. We often dread randomness in systems. Indeed, a key dictate in modern business management is to seek quality by removing all sources of randomness from any process. Given such imperatives, it is easy to think of randomness as a foe to be fought rather than as an opportunity to be embraced. The study of complexity suggests otherwise. Randomness is fundamental to Darwin's theory of evolution, which relies on the notion that errors (variations) during reproduction will provide grist for the mill of selection and result in "endless forms most beautiful and most wonderful."

Darwin's theory, and the role of randomness therein, is really about discovery on rugged landscapes. Our ability to discover new opportunities, whether new forms of animal life or novel technologies, is tied to both the ruggedness of the underlying landscape and our search skills. On simple landscapes, even simple searches can find good outcomes. On rugged landscapes, such searches founder.

Landscapes become more rugged as the elements that compose them interact more. Suppose we are seeking, say, a novel drug cocktail to fight some disease. If each drug we add to the mix has an effect that is independent of the others, then we can quickly find the best cocktail just by adding the drugs one at a time and keeping only the ones that improve the cocktail's overall efficacy. However, if the drugs interact with one another, this simple search strategy breaks down, as the various interactions no longer provide a clear signal on how best to proceed.

It turns out that the introduction of randomness can greatly improve our ability to search on rugged landscapes. As James Joyce noted, "Errors . . . are the portals of discovery." Just as evolution relies on variation to uncover most wonderful forms, introducing errors into a search can be a powerful strategy for discovery.

Accepting randomness in a system forces us to give up some control. Yet when we are facing hard problems, this may be the right thing to do if we want to improve the outcome. More generally, it may be the case that carefully controlled, centralized systems are more of a modern artifact, driven by reductionist thinking, than a universal norm. Indeed, there are plenty of examples where the principles of feedback, heterogeneity, and randomness conspire to create complex systems that are without centralized control, yet quite productive. Effective decentralized decision making may be one of the best new old ideas to emerge from complex systems.

When we think about decision making, our natural tendency is to focus on our own decisions. Over the last few decades entire academic fields have been devoted to understanding how humans make decisions. While unraveling the mysteries of our deciding brain is a worthy enterprise, it is far too easy to overlook the vast number of decisions that take place elsewhere in the biological world. To take just one example, bacteria exist in environments that contain both useful and harmful chemicals, and thus they constantly must make life-and-death decisions about where to move, given the trade-offs among various opportunities. How is this possible without a brain? Even more intriguing, humans (presumably using a brain) and bacteria (presumably not using one) demonstrate similar patterns of choice errors in simple experiments.

The notion that one doesn't need a brain to make good decisions is startling. From the lone bacterium on up to large-scale social systems such as honeybee hives and financial markets, we are surrounded by decision making. How can a swarm of honeybees make good decisions? The queen is not the leader. She leads a rather insular life, serving as a well-tended egg-laying machine, able to emit only signals about her health and existence, rather than operating instructions to the rest of the hive.

Karl von Frisch's discoveries about honeybee communication in the late 1940s inspired generations of scientist to undertake the careful observation and analysis of honeybee behavior. Through this work, we are beginning to understand how a colony can sort out its various options and make good decisions without any central leadership. One particularly important decision for a colony—the difference between its perpetuation and demise—is finding a new location when the old one becomes too crowded.

A swarm of bees solves the problem of finding a new location through the use of a few simple rules and feedback mechanisms. Scout bees, after identifying a potential new site, advertise it to other scouts. The better the site, the more vigorously the scout promotes it. This decentralized process allows the sites to be sorted out and suitably investigated, and ultimately it results in the swarm tending to choose the best site relatively quickly without any central direction.

Understanding such decentralized processes has numerous benefits. It solves an interesting, life-or-death case of honeybee natural history. It also shows how decentralized mechanisms can be used to solve hard problems. This suggests an approach that we might be able to hijack for our own use in, say, coordinating computer networks or large-scale human organizations. Finally, and perhaps most profoundly, such decentralized mechanisms give us new insights into related phenomena. For example, perhaps bees are to neurons as hives are to brains. Are swarm decisions akin to human consciousness?

Complexity arises in systems of interacting agents. Take some agents with simple behavior, connect them together in a particular way, and some global behavior will result. Alter the connections and, often, new global behavior arises. Given this, knowing how patterns of interactions—that is, networks—influence behavior is fundamental to understanding complex systems.

Even in simple models such as lakeside neighbors competing to keep up with one another, interesting patterns begin to emerge. Starting from such a simple system, we can alter the connections slightly and find radically different behavior taking over. Indeed, by introducing only a few long-range connections, we find that it may be a small world after all, where anyone can connect to anyone else using only a few intermediaries. If neighbors can connect to one another, they can influence one another. Thus, the networks that define neighborhoods drive system-wide behavior.

This behavior is often surprising. For example, a well-mixed world where neighbors are tolerant of others easily segregates into neighborhoods of homogeneous types.

One of the more surprising principles coming out of the complexity that abounds is the existence of scaling laws. Starting in the late 1800s, biologists began to notice that, when appropriately scaled, various physical and physiological features of a variety of organisms aligned in a simple way. A simple rule links the metabolism of a single cell to that of a blue whale. Knowing the heart rate and weight of, say, a mouse allows us to predict the heart rate of, say, a thousand-pound cow. The ability to make such predictions is tied to the fundamental constraints that govern such complex systems. In this case, limits on how densely we can pack the pathways needed to provide resources to the organism drive the scaling.

Scaling laws arise in other complex systems as well. The size of cities or firms tends to follow well-defined laws, with the largest having twice the size of the second-largest, three times that of the third-largest, and so on. Similarly, in a book, the word that is most commonly used is twice as likely to occur as the next most commonly used word. Even the number and death tolls of wars are governed by a scaling law.

Knowing the scaling laws that govern our lives provides a portal into our future. For example, over the last century we have seen a trend toward urbanization. More than half of the world's population now lives in urban areas.

Is such a trend good or bad for humanity? The answer to this question is tied to the coefficients of various scaling laws of cities. These will tell us whether more urbanization will allow us to use fewer resources, be more inventive, and so on. Similarly, the scaling laws of wars may hint at how many conflicts with how many deaths we are likely to see in the future.

In complex social systems we often see the emergence of cooperation. Agents in systems can either compete or co-operate with one another. Competition makes you slightly better off, while cooperation makes you much better off. Unfortunately, most social systems have incentives that favor, at least individually, competition over cooperation. Such systems can easily end up with the inferior, competitive outcome.

Notwithstanding incentives to compete rather than co-operate, complex social systems may find ways to achieve the cooperative and socially superior outcome. On the island of Bali, farmers have been farming the picturesque rice terraces sustainably for more than a thousand years. This cooperation persists despite what would appear to be overwhelming economic incentives to compete with one another for the scarce water. However, by carefully unraveling the complex dynamics that govern this ecosystem and applying the principles of feedback and networks discussed above, we can resolve this apparent anomaly. Oddly, the neighborhood feedbacks from the presence of damaging crop pests and diseases realign each farmer's incentive to share water, and with such sharing, society is better off. Moreover, the newfound need for coordinated cropping opens up a niche for an elaborate religious institution with various shrines and temples tied to the irrigation systems.

We can also formulate an abstract model from which we can observe and understand the emergence and persistence of cooperation. We find that in a world red in tooth and claw, where competition can easily overwhelm the system, slight variations in competitive strategies provide a means by which cooperation can emerge. Cooperative agents develop a way to communicate so as to recognize one another. By doing so, they get the benefits of cooperation while minimizing losses when they encounter competitive agents. Through such a mechanism, cooperation can emerge and be sustained.

The final principle we will explore is that of self-organized criticality. Consider grains of sand being slowly piled on a table. As we drop each grain, it might land on a stable spot and increase the pile, but in doing so it makes that spot less stable than it was before. Alternatively, it might land on an unstable part of the pile, triggering an avalanche.

Over time, this interplay of stability and instability self-organizes the sand pile into a critical state. Once this happens, we find that avalanches of all sizes are possible (the distribution of which is described by a scaling law), with smaller ones far more likely than larger ones.

One implication of the sand pile is that once we enter the critical regime, the dropping of a single grain of sand can cause, on rare occasions, an avalanche encompassing the

entire pile. Various social systems may evolve toward similar
critical states. We might find ourselves in a world governed
by the mathematics of the sand pile. Stock markets may be
subject to numerous routine adjustments as typical world
events transpire. Yet these same types of events will, on rare
occasions, lead to a massive readjustment. Civilizations may
be governed by political systems that tend to push people
toward critical states, where small events occasionally result
in the collapse of an ancient civilization or, as we saw in the
Arab Spring, modern governments.

We will conclude our exploration of complexity
by following an arc that begins with our desire to understand atomic interactions at the start of the atomic and
information ages and ends with a new fundamental theorem about complex adaptive systems. In the early 1950s
Nicholas Metropolis and others developed an algorithm to
explore interacting molecular systems. At the core of this
algorithm is a set of simple manipulations that ultimately
allows one to recover a critical piece of information that
is impossible to generate directly; as if by magic, this algorithm produces the unknowable. Related algorithms have
become a critical component in our emerging analytic age,
as they solve a key problem in applying the eighteenth-
century statistical ideas of Presbyterian minister Thomas
Bayes to real-world problems ranging from targeted web
advertisements to driverless cars.

At the heart of complex adaptive systems are agents searching for better outcomes. With a few simplifications, the key aspects of this search behavior can be linked to elements of the algorithm above. Thus, agents in such systems are, unknowingly, performing a dance governed by a cosmic algorithm. Given this connection, we derive a new theorem of complex adaptive systems that embraces the magic inherent in the algorithm. This new theorem implies that as agents adapt in these complex systems, their adaptations are governed by probabilities tied to their underlying fitness. While agents are more likely to be found concentrating on the better solutions, there is always a (lower) chance that they will find themselves in suboptimal circumstances. This is a result that is at once both gratifying and humbling, as it suggests that while agents will often find the best outcomes, they will inevitably fail on occasion.

Complexity abounds. Exploring its core principles will take us on a journey across the scientific landscapes outlined above. It is a journey marked by awe, inspiration, and ultimately insights that are critical to our scientific understanding of the world around us and to our ability to survive when confronted by our most challenging problems. It is a journey about true places, where the maps are not always well formed, but they are suggestive enough to be of use given our innate desire and need to explore this frontier.

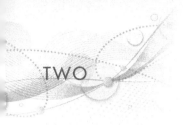

TWO

From So Simple a Beginning:
Interactions

> There is grandeur in this view of life, with
> its several powers, having been originally
> breathed into a few forms or into one; and
> that, whilst this planet has gone cycling on
> according to the fixed law of gravity, from
> so simple a beginning endless forms most
> beautiful and most wonderful have been,
> and are being, evolved.
>
> —CHARLES DARWIN, *The Origin
> of Species*

WE ARE SURROUNDED BY "ENDLESS FORMS MOST BEAUTIFUL
and most wonderful," whether they are embodied by one of
the myriad of species we find on our planet or in artificial
structures such as the New York Stock Exchange. Darwin's
keen insight, the one that put grandeur in his view of life,

was that reproduction with variable inheritance and natural selection could move us from simple beginnings to extraordinary ends. A related insight, first postulated by physicist Phil Anderson in 1972, sits at the heart of complex systems. It holds that simple pieces, interacting together, can result in the emergence of new, most wonderful forms.

The potential emergence of new forms from the aggregation of simple pieces is known as the "more is different" hypothesis. This hypothesis is a direct challenge to the foundations of modern science.

At the core of modern science is a belief in the power of reductionism: the idea that to understand the world we only need to understand its pieces. Thus, if we can fully understand atoms, we will then understand chemistry, as chemistry *just* studies collections of atoms, and from there we will know biology, as it relies on chemistry, and on and on. Similarly, in social systems, if we can understand a neuron, we will understand the brain, and thus know individual decision making, which allows us to understand group decision making, which gives us deep knowledge of governments and firms, and ultimately a full understanding of economics, politics, and society at large.

The key insight from the "more is different" hypothesis is that reductionism does not imply constructionism. That is, even if we can study and know the world's simplest components, that does not imply that we will understand everything just because the world is constructed from these components. Indeed, to reconstruct the world we have to have a theory of how components, once put together, interact. There is an old Sufi adage that even if you understand the number 1 and know that 1 *and* 1 make 2, you don't understand 2 until you know what "and" means.

Take the letters on this page. Each letter is composed of a few hundred dots per inch in careful relation to one another. Yet there is something inherent in these dots that allows letters to emerge, even if the relationship of the dots is somewhat altered, as you might see in the distorted image from a CAPTCHA challenge found on some web pages (see Figure 2.1).

Moreover, the letters, when placed near one another, take on new properties and meaning, and ultimately result in the emergence of words. Such emergence is so strong that it persists even if we scramble the letters between each word's start and end, that is, such eregmnece is so stnorg taht it pesstirs eevn if we sabcmlre the ltteres betewen ecah wrod's satrt and end.

Illustrating the above ideas is a branch of mathematics—initiated by the freakishly accomplished John von Neumann—that studies structures called cellular automata.

FIGURE 2.1: The distorted letters of a Completely Automated Public Turing Test to Tell Computers and Humans Apart (CAPTCHA) rely on the complexity inherent in the emergence of letter forms within the human mind.

Start with an empty checkerboard, and across the top row randomly place some checkers. For each subsequent row, we will place a checker in a particular square based on the pattern of occupied squares in the row above and some fixed rule. For example, suppose the rule is that you only place a checker on a square if the square immediately above it is occupied. If we dutifully follow this rule, each new row will duplicate the row above, and slowly our checkerboard will be filled with vertical stripes located wherever we happened to have placed a random checker in the top row. Obviously, this rule is rather boring, though it does hint at how a simple, very localized rule (it just looks at the square immediately above and ignores squares that are more distant) can result in a global pattern, in this case a set of vertical stripes that, if you are willing to be indulgent, resemble the stripes of a zebra.

Let's add a bit of complication to the rule. Suppose our rule depends not only on the square above but also on that square's immediate left- and right-side neighbors. There are 256 possible rules of this type, and with some malice afore-thought, let's use one of the following form: if only one of the three squares above is occupied or if only the square above and its right-side neighbor are occupied, then add a checker, otherwise leave it empty (for aficionados of this genre, this is known as Rule 30). Figure 2.2 illustrates one of the possible patterns that can emerge from this rule. Notice how the pattern has a lovely theme of inverted triangles of various sizes being placed at seemingly random locations.

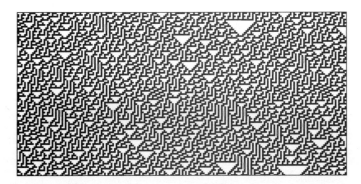

FIGURE 2.2: A pattern emerging from a cellular automaton using Rule 30 and random initial conditions. Here there are a large number of cells across the lattice and the left side wraps around to the right side, forming a cylinder. *(Generated by WolframAlpha.)*

Furthermore, note how some of these structures extend across many squares of the checkerboard. Such large-scale structures are surprising, given that any individual square uses information only from the three squares immediately above it, yet the structures that form span tens of squares rather than just triplets.

Our exploration above provides a proof of the concept that interesting (and perhaps even complex) global patterns can emerge from simple local rules. Of course, knowing that something is possible doesn't mean that it exists in nature or, even if it does, that it is important. However, in this case, such behavior appears to be an important part of our world.

Consider a cone snail, a seemingly lowly species of sea snail. There are at least two surprises that make cone snails remarkable. The first is that they are potentially

FIGURE 2.3: The shell pattern on a *Conus omaria* sea snail. *(Photograph by the author.)*

deadly, as they have a harpoon-like tooth (which comes out of the pointy end with surprising speed and dexterity) that is attached to a poison gland that contains some very effective neurotoxins. (Beware the escargot!) The second surprise, relevant to this discussion, is that the outside of the shell of some species is beautifully patterned, as seen in Figure 2.3.

What makes these cone snail shell patterns particularly intriguing is their similarity to the patterns that emerge in the cellular automaton discussed above. We are not claiming that cone snails use Rule 30 to pattern their outside, but only that it is possible that some local rule, rather than some global plan, is responsible for what we see.

Indeed, could it be any other way? At one extreme we could think of a global plan for the pattern of the shell. As the shell grows, the snail knows what goes where based on the encoded master plan and directs the construction accordingly. One could even invoke an intelligent designer for the resulting pattern, perhaps to give the shell some camouflage for hunting. Alas, such explanations seem rather superfluous given a much simpler alternative.

The snail's shell grows by accretion at its edge. As it adds new material, the pigmentation is determined by various chemical processes of activation and inhibition that by physical necessity are tied to local conditions. Thus, a natural rule along the lines of something like "If there is only one dark-pigmented cell in your neighborhood, accrete a dark-pigmented cell, otherwise accrete a light-colored cell" (perhaps because too many dark cells inhibit the formation of new ones and too many light ones activate the formation of a dark one) gets us almost to Rule 30. Of course, Rule 30 also differentiates between right- and left-side neighbors. However, such asymmetry (known as chirality) arises in biological systems as well. For example, in seashells, almost all individuals in any given species have shells that coil in the same direction.

Given the above two options about how the pattern arises on the outside of a cone snail shell—a master pigmentation plan that is maintained by the snail with additions to the shell being carefully monitored and directed, or one where very localized chemical interactions determine the pigmentation of added cells—it is not hard to favor the simpler explanation. The only non-intuitive part of such a hypothesis is that the patterns that emerge, with their riffs on the inverted-triangle theme, seem a bit too clever to be generated by such local means. If not for the existence proof provided by the cellular automata above, we might not believe that such a thing was possible.

The notion that local interactions can result in interesting global patterns has some important implications for evolution. Indeed, a new branch of evolutionary science that focuses on the relationship between evolution and the developmental processes of organisms, called evo-devo for short, embraces this perspective.

Consider the *purpose* of the pattern that we see on the cone snail shell. In a world driven by evolution, that pattern is more likely associated with the more successful cone snails, either because it provides some direct fitness advantage or because it takes a free ride on something that does. In this case, the pattern likely helps our slow-moving yet carnivorous sea snail either by providing it some camouflage from, or by making it attractive to, its prey.

In the checkerboard automata above, we saw how a simple rule can generate a global pattern. Indeed, for automata that only depend on the square above and its immediate neighbors, we need just eight bits of information to define a rule. (There are eight possible configurations of three contiguous squares, and we need one bit of information to determine whether we place a checker in the lower square given each configuration.) By making a minor change to one of these eight bits, we will get a new rule that most likely generates an entirely new pattern. For example, in Figure 2.4 we took Rule 30 and made a minor change, namely,

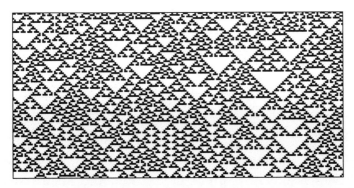

FIGURE 2.4: A pattern emerging from a cellular automaton using Rule 22, a "one-mutant" neighbor of Rule 30, and random initial conditions. (*Generated by WolframAlpha.*)

placing a checker only if there is exactly one checker in the three neighbors above (recall that Rule 30 did this, but it also placed a checker if the parent and right-side neighbor were both occupied).

This new rule is known as Rule 22. As you can see, it results in a more regular and symmetric pattern, which is probably bad if you want to camouflage yourself. On the other hand, this new pattern seems much bolder, and the additional symmetry induces some curiosity in the viewer (at least in this viewer). Thus, by a slight alteration to the underlying chemistry (rule), the development process results in a very different pattern that potentially could benefit the cone snail—especially if feeding on curious humans drawn by the additional symmetry proves to be an advantage. Regardless of this particular example, the core idea, that even small changes to local processes can have big implications, is important: from so simple a beginning endless forms . . .

Like any good model, cellular automata are able to capture the essence of an important phenomenon in a very exact and sparse way. By doing so, they provide a constructive proof of how simple local rules can have complex global implications. There is an alternative form of complexity that is also of interest, namely, systems where complex local behavior results in a simple global outcome. Such systems were first formally discussed more than two hundred years ago, and they form the basis of our modern economy.

In 1776, Adam Smith published *An Inquiry into the Nature and Causes of the Wealth of Nations*. In a brief passage within this tome he wrote, "He intends only his own gain, and he is in this, as in many other cases, led by an invisible hand to promote an end which was no part of his intention." Almost two hundred years later, in 1954, Ken Arrow and Gerard Debreu provided a formal existence proof of Smith's hypothesis, namely, that under certain conditions one can find a set of prices under which economic agents—each out for her own gain—will want to buy or sell just enough of each commodity to equilibrate prices and maximize society's gains from trade. Thus, the apparent chaos of the marketplace is replaced fortuitously by a grand clockwork, not part of anyone's intention, that brings everything into balance, and even results in a set of trades whereby we cannot improve anyone's lot in life without harming someone else. As Smith put it: "By pursuing his own interest he frequently promotes that of the society more effectually than when he really intends to promote it. I have never known much good done by those who affected to trade for the public good."

Smith's insights into what economists call general equilibrium are part of a remarkable arc of economic thinking and intellectual triumph spanning hundreds of years. Such thinking is often driven by trying to resolve a paradox. For example, water, an essential element for life, is cheap, while diamonds, an inessential bauble, are expensive. How can this be?

The answer to such a conundrum is that considering only the need for a good—the so-called demand side of the market—gives us only part of the picture. Along with the demand for a good, one must also consider its supply. Thus, drinkable water is abundant (though this is changing), while diamonds are rare and (somewhat) difficult to find. The medieval Muslim economist Ibn Taymiyyah wrote in the 1300s, "If desire for the good increases while its availability decreases, its price rises. On the other hand, if availability of the good increases and the desire for it decreases, the price comes down." Such thinking gets refined in subsequent generations by thinkers such as John Locke in 1691 and James Denham-Steuart (who first used the phrase "supply and demand" in a book published in 1767, just nine years before Smith's *Wealth of Nations*).

In 1870, Fleeming Jenkin published a paper, "On the Graphical Representation of Supply and Demand and Their

Application to Labour," that illustrated the power of the supply and demand graphs that—after some refinements and popularization by Alfred Marshall in 1890—every budding economist learns in her introductory class. The supply and demand diagram is one of those rare and remarkable scientific illustrations that takes a complex reality and summarizes it in a simple, valuable way. (Other such diagrams range from the commonplace, such as the high- and low-pressure fronts shown in the daily weather map, to the exotic, such as Feynman diagrams, used to track the contributions of particular particle classes in quantum field theory.)

The basic ideas behind supply and demand are straightforward (at least in hindsight). First, we separately consider the behavior of the potential suppliers and demanders of the good in response to various price changes. For example, suppose we have two suppliers who can produce a good at a cost of $10 and two who can do it for $30. We can summarize the behavior of these four suppliers by drawing a supply curve that shows how many goods will be offered for sale (on the x-axis) under various prices (on the y-axis), as shown in Figure 2.5. (Alas, the axis designations used here are a historical artifact at odds with the usual scientific convention of listing the independent variable—here, the price—on the x-axis.) Thus, at a price of $5, no one is willing to sell, while at a price of $25, both of the suppliers with costs of $10 will offer their goods to the market, while the two suppliers with costs of $30 will abstain, and so on. Similarly, we can summarize the demand

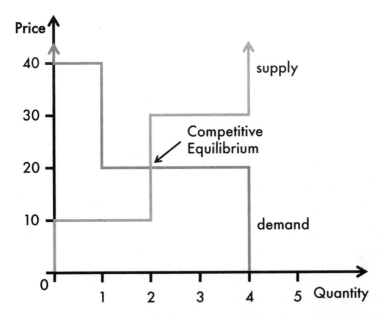

FIGURE 2.5: Supply and demand in a simple market. This market has two sellers with production costs of $10 and two with costs of $30, and one demander with a value of $40 and three with values of $20. In competitive equilibrium we would expect two goods to be traded at a price of $20. These trades would be among the two sellers with production costs of $10 trading with the one demander with a value of $40 and one of the three demanders with values of $20.

side of the market by graphing the potential actions of the demanders. Suppose we have three demanders who would be willing to pay up to $20 for a good, and one that is willing to pay $40. Then, at a price of $30, only the demander with the $40 value for the good will want to buy, while at a price of $20 or below, all four demanders will want to purchase a good.

The supply and demand diagram represents a nifty summary of a remarkable amount of information. Neatly contained within each of the curves are the inherent forces that drive the market. The supply curve captures the current production technology, the motivation of workers, the availability of the input goods that must be transformed into the final product, and so on. The demand curve captures the desires that individuals have for the good, the availability of alternative goods, and other such factors.

Knowing the shape of the supply and demand curves alone is like knowing where the high- and low-pressure areas are on a weather map—somewhat interesting, but useful only if you have some theory for what happens when the two fronts interact.

Economists typically rely on the notion that systems will seek out an equilibrium, and from this principle, they predict what will happen. Of course, there is nothing inherent in our world that suggests systems equilibrate, but such an assumption does have a few advantages. First, there may be systems where external forces do indeed tend to push the system to a state of rest. For example, consider dropping a ball into a bowl—the gravitational force will cause the ball to roll downhill, and it will eventually come to rest at the lowest point of the bowl (or, if the bowl is badly dented, at the bottom of one of the dents). Moreover, if you give the ball a slight push from its resting place, the forces will conspire to move it back to its original position. Of course, not all equilibria are so stable. For example, if we invert the bowl and carefully balance the ball on top, the ball will remain there, but even a slight whiff of air will have the ball heading off the bowl and settling somewhere far away. The second advantage of seeking out an equilibrium is that it tends to make the analysis much easier (though sometimes an equilibrium is easy to recognize once you find it but hard to find in the first place, as in the case of a combination on a safe). Finally, an emphasis on equilibrium is innately comforting, as it is much nicer to think of a system as part of a grand clockwork aligning the world into a finely balanced state, rather than as something undergoing random wanderings.

In markets, economists have a notion of competitive equilibrium. The idea is simple: markets should equilibrate at a price where the amount that the suppliers want to sell just equals the amount that the demanders want to buy. Thus, by announcing such a price to the market, the desire for goods will just equal their availability, and every supplier who wants to sell can find a demander who wants to buy, and equilibrium will ensue. In our previous graph of supply and demand, competitive equilibrium occurs at a price of $20. At this price, the two $10 suppliers want to sell, while the two $30 suppliers do not. At the same price, the $40 demander wants to buy, while the remaining three $20 demanders are indifferent, being equally well off whether they buy or not. Thus, the theory of competitive equilibrium predicts that only one of these remaining three demanders will come onto the market and buy, resulting in an equilibrium with a price of $20 and two goods sold.

There is, surprisingly, a more subtle result embedded in competitive equilibrium. If you think about the *total* amount of profit earned by the traders in the equilibrated market, you will see that they walk away with $40 (the high-valued demander paid $20 for something she valued at $40, so she earns $20, and the two suppliers each earn $10 by trading at the equilibrium price of $20). Is there some way to rearrange trades that would increase the total amount earned? Suppose that the high-valued demander traded with one of the sellers with a cost of $30, generating $10 of profit between the two of them (with the share of the profits going to each depending on the specific price they agree upon). This would leave the two $10 suppliers to deal with the three remaining $20 demanders, so two additional trades would result, each generating a profit of $10. Here, the total amount of profit earned by the three trades is $30, which is $10 less than what we had before. Indeed, one can show that trading patterns other than the one resulting from competitive equilibrium will only reduce the amount of total profit earned by all of the traders.

The focus on maximizing the total profit from trade is important, since if we are not maximizing this profit, we are losing out on an opportunity to make at least one person better off without harming anyone else (assuming our traders only care about their individual profits). To see this, suppose that the market outcome is inefficient in the sense that the resulting set of trades does not maximize total profit. Whatever total profit was earned in this inefficient market gets divided among the traders somehow. Now, rerun the market and maximize total profits. Since the maximized profits are greater than the profits from the inefficient market, this time we have enough profit to give every trader exactly what she earned in the inefficient market and still have some profit left over. We can then take this leftover profit and give it to one (or more) of the traders, making the recipient(s) better off without harming anyone else. This latter insight, plus the idea that the traders are out for themselves, brings us full circle back to what Smith wrote: "He is in this, as in many other cases, led by an invisible hand to promote an end which was no part of his intention."

The complex-systems view of markets differs from the above account. Recall that to equilibrate the market, we first announced the competitive equilibrium price, and from there, all was well. But where did that price come from? The market is composed of individual suppliers and demanders, each of whom knows only her own cost of selling or value of buying. Given this, how does the competitive equilibrium price ever emerge? As Friedrich Hayek—an early proponent of the complex-systems perspective—made clear in 1945:

> The problem is thus in no way solved if we can show that all the facts, if they were known to a single mind (as we hypothetically assume them to be given to the observing

economist), would uniquely determine the solution; instead we must show how a solution is produced by the interactions of people each of whom possesses only partial knowledge. To assume all the knowledge to be given to a single mind in the same manner in which we assume it to be given to us as the explaining economists is to assume the problem away and to disregard everything that is important and significant in the real world.

Answering Hayek's challenge may not be as hopeless as it appears at first blush, as we may be able to hypothesize some market mechanism—for example, an auctioneer who stands up in front of everyone and announces potential prices, getting a sense of how many suppliers and demanders want to trade at each price, and from this tedious exercise she can generate the competitive equilibrium price. Of course, we see no such auctioneers in the real world. Instead we see various auction institutions such as specialists in the New York Stock Exchange or the colorfully jacketed traders yelling and gesturing to one another in the commodity pits of Chicago.

Unfortunately, it has been extremely difficult to derive a well-grounded theory of how prices arise in markets. While the theory of competitive equilibrium is innately compelling given that any imbalances in supply and demand should push prices in a way that will bring trades in line, it is hard to imagine how such forces are actually directed in the real world.

Can we build an alternative complex-systems theory of markets from the bottom up? That is, can we make some simple assumptions about trading and, from these, show how global patterns of trades and prices emerge?

My colleague Michele Tumminello and I have been pursuing this approach by considering a simple trading bazaar. In this bazaar, traders wander around and bump into one another. When they meet, traders blurt out a random offer, with the only proviso being that if that offer is accepted, the trader will not lose money. To make this conceptually simpler, let's just assume that if two traders bump into each other and there is the possibility of a mutually profitable trade, they will trade at a price given by the midpoint between the supplier's cost and the demander's value (it is an easy extension to make the traders a bit more coy). If two traders cannot agree on a deal, they continue to wander around and bump into new potential trading partners.

In such a bazaar, using the same set of suppliers and demanders discussed above, there are two possible configurations of trades that can arise. The first, closely related to that arising under competitive equilibrium, has the $40 demander trading with a $10 supplier (at a price of $25) and one of the $20 demanders trading with the other $10 supplier (at a price of $15). Note that while these are the same traders that are involved in transactions under competitive equilibrium, the prices differ, since competitive equilibrium predicts that both trades will occur at a price of $20, versus the $25 and $15 predicted here.

An alternative trading configuration is also possible. Suppose instead that the $40 demander initially bumps into one of the $30 suppliers. In this case, they will agree to trade at a price of $35. At this point, the only remaining pairings that can result in mutually agreeable trades are between the $20 demanders and $10 suppliers, so we would predict that two such trades will occur eventually (since we have three such demanders but only two such suppliers) at a price of $15. Thus, if history unfolds as in this scenario, we get three trades, one at a price of $35 and two at a price of $15. This configuration is quite different from what we would predict under competitive equilibrium, and it is inefficient in the sense that the total profit earned across all of the traders is only $30, versus the $40 we get under competitive equilibrium. Thus, the system as a whole lost out on $10 of additional profit that could have been used to improve at least one person's lot.

While it is somewhat tedious (without a computer, especially for big systems), one can work out the likelihood of the two configurations above arising in the bazaar. About one-third of the time (8/25, to be exact), we will get the configuration associated with the competitive equilibrium outcome, and the alternative configuration will arise around two-thirds of the time (17/25).

Thus, under the bazaar model—with the potential for easy wordplay fully acknowledged—there is roughly a one-third chance that we will end up with the same trades that we see under competitive equilibrium, though at slightly different prices. The remaining two-thirds of the time we expect to see a very different outcome, with one trade at $35 and two at $15. In both worlds, traders acting only in their own interests produce an outcome that was no part of anyone's intention, namely, a set of prices and a pattern of trades that result in some aggregate profit across society.

Given the above two models, which one should we believe? This is a difficult question. If we look at data (see Figure 2.6) generated in experimental markets (similar to the one above but with many more traders), we see peaks in the data that correspond to the midpoint values we predict from the bazaar model, rather than the uniform price predicted by competitive equilibrium. Of course, with any model of behavior, we allow some room for error. Thus, having prices that don't quite line up is expected, and one needs to make a judgment about whether the misaligned prices are closer to those emerging from the competitive equilibrium model or those from the bazaar model. Given our data, it seems that the bazaar model cannot easily be dismissed.

Different views of the same phenomenon are often useful in gaining a deeper understanding of a system. In many ways, the competitive equilibrium and bazaar models complement each other, improving our understanding of markets. Of course, academic paradigms often solidify around a particular view, and the ability to alter this point of view depends on the field. In physics, for example, simple models that better explain the data tend to quickly win the day. In

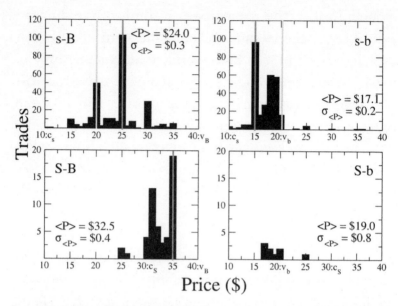

FIGURE 2.6: Price distributions arising from some experimental markets with human subjects. Each market had multiple subjects, with demander values of $20 and $40, and seller production costs of $10 and $30. Each panel of the figure shows the distribution of observed trading prices between buyers and sellers with the indicated values (s for seller costs of $10, S for seller costs of $30, b for buyer values of $20, and B for buyer values of $40). Competitive equilibrium predicts that all trades will occur at a price of $20 (indicated by the light vertical lines in those distributions where trades are predicted) and that the underlying buyer value and seller cost in each trade will make no difference in the distribution of observed prices. The bazaar model predicts trades at the intermediate price between the buyer values and seller costs (indicated by the dark vertical lines in those distributions where trades are predicted).

economics, accepting new approaches that may better explain the data is a much slower process. Economists ignored experimental data until relatively recently and typically embrace a very prescribed modeling paradigm that relies on optimization and equilibrium. Ultimately, we must follow Kenneth Boulding's first law, namely, "Anything that exists is possible"—an obvious and useful observation that often gets ignored.

Whether the competitive equilibrium or bazaar model is the right description of our world is an open question. Experimental markets with naive traders, like those shown in Figure 2.6, certainly seem to carry the signature of a trading bazaar rather than a carefully orchestrated market. Indeed, one might put more faith in the competitive equilibrium model if our traders found themselves in the same market day after day, or if we introduce some additional rules of engagement—for example, perhaps bids and offers must be clearly posted for all to see and for anyone to accept. Such additional conditions might push the outcome of the market more toward the predictions of competitive equilibrium.

Regardless, both models are interesting in that they embrace the fundamental notion of complex systems: that interactions among individuals can result in the emergence of global outcomes—in this case, patterns of prices and trades—that were no part of anyone's intention.

For more than one hundred years, economists have been relying on the competitive equilibrium model to predict how markets will behave and, in turn, how policies should be made. The competitive equilibrium model and the tools of supply and demand are a real triumph of science. Consider again the fundamental problem we face when analyzing a market. Potential traders, knowing only their own values and costs, come together to try to make self-interested deals with one another. These potential traders, some bored and tired, others motivated and sharp, randomly get together and try to close a profitable deal as they gather pieces of information from the shouts and murmurs of their fellow traders along the way. Perhaps some traders are calculating thinkers who derive the best trading rules possible given the limited information they face, while others are more reckless and trade willy-nilly.

Out of such chaos comes order, in the form of a stream of trades and prices. The idea of competitive equilibrium is an extreme version of such order, where the cacophony of shouts results in a single price sufficient to balance everyone's desire to trade, with just enough goods being offered by the sellers to meet the buyers' demands, and the resulting trades maximizing the total profit available in the market. In such a model, displacing even one of these trades would result in society having less.

An alternative tale of order is that of the trading bazaar. In this world, we dismiss the notion that a unique, global, competitive equilibrium price emerges, and we embrace the chaotic machinations of the traders. Potential buyers randomly encounter potential sellers, and seemingly randomly generated offers are proffered. When an offer results in a mutually profitable trade, it is accepted and the traders leave the market. Again, a predictable (though here with a bit less certainty) set of prices and trades emerges in the market. The result is certainly messier than that arising in competitive equilibrium, and formal tests of the performance and utility of the two approaches are only now being conducted.

The power of local interactions to form unexpected global patterns is remarkable. Whether these local interactions lead to the beautiful design on the shell of a sea snail or to a set of prices and trades that maximize society's bounty, from such simple beginnings wonderful forms are emerging.

THREE

From Flash Crashes to Economic Meltdowns: *Feedback*

> Without calling the overall national issue
> a bubble, it's pretty clear that it's an
> unsustainable underlying pattern.
>
> —ALAN GREENSPAN

O N THURSDAY, MAY 6, 2010, AT 2:32 P.M. EASTERN STANDARD Time, a sequence of events began that led to chaos in the securities markets for the next half an hour. During the first few minutes of this period, major United States equity indices plummeted 5–6 percent (see Figure 3.1). At 2:45 p.m. a five-second trading pause was imposed on the market, and the indices miraculously rebounded. However, the tsunami that started in the indices began to wash over the entire equities market. Trading prices for more than three hundred individual stocks deviated by

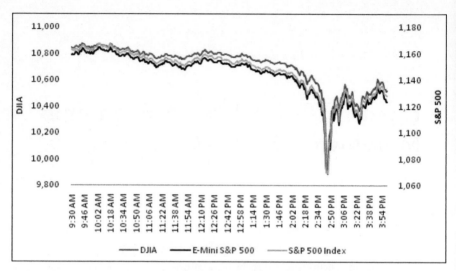

FIGURE 3.1: Major US market indices on May 6, 2010, including the Dow Jones Industrial Average (DJIA, left-hand scale) and the Standard and Poor's 500 Index (S&P 500 Index, right-hand scale). *(Source: US Securities and Exchange Commission.)*

more than 60 percent of their previous values. As liquidity dried up in some stocks, markets failed, and at the extreme the prices of shares in formerly well-regarded companies began to fluctuate wildly, with a company such as Accenture, previously trading at $40, selling for only a penny, and shares of Apple quickly moving from $250 to $100,000. By 3:00 p.m. the tsunami had subsided, and the markets returned to more normal behavior.

What could have possibly started such turmoil? Was there some news report of a cataclysmic event, like a major war breaking out or the assassination of a key world leader? Did some European country suddenly default on its debt?

Was there some terrorist strike on the homeland or cyber-attack on the trading system? Alas, the triggering event was at once far more mundane, and far more worrisome, than any of these.

The proximate cause of the above turmoil appears to have been a set of trades initiated by a money-managing firm whose address was a post office box in Shawnee Mission, Kansas. This firm used a computerized trading program to sell some securities, relying on an algorithm that tied its trading behavior only to the current *volume* of trades on the market, rather than to a more obvious factor such as the security's price. While in hindsight it is easy to see how such a program could induce a ripple in the market waters, the far greater concern is how the ever-growing set of interconnections and interactions across the complex financial system was sufficient to allow this ripple to grow into a full-fledged tsunami that, at least for half an hour, wreaked havoc upon the financial shores.

In September 2010, the US Commodity Futures Trading Commission and the US Securities and Exchange Commission released a joint report on the aptly named "flash crash," entitled *Findings Regarding the Market Events of May 6, 2010*, and many of the market details below are drawn from this source. The report provides a detailed economic autopsy of the events that unfolded on that day, and it's rather good reading (and easily available by download), all things considered, for those inclined to explore the intimate details. Like any good autopsy, contained within its dry descriptions and careful analyses is a remarkable story of how the death came about. Yet the real intrigue comes from what is not discussed, namely, the mystery about who did it, why, and whether it could have been avoided.

May 6, 2010, had begun with the financial markets already on edge. The European debt crisis dominated the political and economic landscape, especially the possibility that Greece might default on its debts. Changes in various market indicators, such as the expected stock market volatility, the premium on debt insurance, the exchange rate of the euro, and the prices of gold and safe securities such as Treasuries, all reflected the unease brought about by these conditions. These changes likely pushed the markets toward a critical state (an idea that we will explore in Chapter 11), where even a small event had the potential to cascade into a much larger chain reaction.

The start of the tsunami began innocently enough. A firm that managed mutual funds wanted to hedge its existing equity positions against future changes in the US equities market. This is a common desire, as presumably the firm wanted to lock in some profits it had recently gained against the possibility that the US equities market might decline. To accomplish this hedge, the firm wanted to sell 75,000 E-mini futures contracts that would become due that June. E-minis are a derivative security, that is, their value is tied to something else—in this case, each one is worth fifty times the value of the S&P 500 Index (which represents about 70 percent of the market capitalization of all US-listed equities). Thus, if the S&P 500 Index is at $1,000, each E-mini is worth fifty times that, or $50,000. The 75,000 contracts represented roughly 3.4 percent of the average daily volume traded during 2010, and they were worth a total of about $4.1 billion at the prevailing price. Having a single individual wanting to sell 75,000 contracts is unusual, although there had been two occasions in the preceding twelve months where trades of this size or larger had been conducted.

Such trades are not without hazards. The problem with selling such a large number of shares all at once is that you can easily cause the price to plummet if you are not careful. Suppose you decide to sell a large number of shares by just dumping them all on the market at once. Initially, if the market is liquid, there are some buyers around to purchase your shares at roughly the going price. As these buyers' demands are satisfied they leave the market, and your shares begin to flow to preexisting offers to buy that have been recorded by the exchange in its "order book." As your shares satisfy the highest of these preexisting offers, they next flow to lower ones, and on and on. As your orders eat well into the order book, prices continue to fall, and any potential new buyers coming onto the market recognize what is happening and make much less aggressive offers in anticipation of lower prices, putting even more downward pressure on prices. While any large offer to sell will have a tendency to depress the overall price regardless, given the microdynamics of the order book, dumping the shares on the market all at once will have a larger, short-lived impact, resulting in the seller getting much worse prices overall than she could have gotten if the shares were sold more slowly.

Thus, to get the best prices possible for a large order, the seller needs to carefully manage the block of trades and slowly release shares onto the market. This allows new buyers to find their way to the market during the sale, refilling the order book in the process, and ultimately resulting in the seller getting much higher prices overall for the entire lot.

One way to manage the sale of a large block of trades is to employ an automated trading algorithm that executes the orders in a reasonable manner. Such an algorithm should be programmed to track key data from the market, such as the current trading volume, price, and time of day. Based on this information, the computer will release the shares so as to get the best deals possible consistent with moving the entire block in a timely manner.

The firm that catalyzed the flash crash used such an algorithm. Of course, the devil is in the details, and in this case there was a devil indeed. The firm's algorithm had one simple rule: feed in orders to the market so that these orders constitute less than 9 percent of the overall trading volume during the previous minute. Note that this algorithm completely ignores the trading price. That being said, at some level it is not a completely absurd algorithm, as normally volume is a good indicator of the market's liquidity, and liquidity is tied to stable prices. In theory, if you remain a small part of the market (here, less than 9 percent) and the market is functioning in a "normal" way, this algorithm should result in stable and reasonable prices. In essence, the algorithm hitches a free ride on the volume information coming from the market and uses this as a proxy for reasonable prices. By doing so, it avoids having to make any difficult predictions about when to sell.

Unfortunately, there have been two recent changes in trading that have made this volume-based proxy quite dangerous. First, the rise of derivative securities has interconnected various markets. E-minis are linked to the S&P 500 Index. There are other derivatives with slightly different designs, such as S&P Depositary Receipts (known as SPDRs or "Spiders" and traded under the ticker symbol SPY), that are also tied to this index. If the price of one of these derivatives differs substantially from the others, there is an arbitrage opportunity to lock in a profit, regardless of what happens to the underlying prices, by selling the more expensive security and buying the cheaper one to make good on the previous sale. An alternative arbitrage opportunity interconnects the derivatives market to the broader markets: given that the price of the derivatives is tied to a bundle of individual stocks (which make up the index), you can always make a profit by offsetting the purchase (or sale) of the derivative by selling (or buying) the underlying bundle of stocks whenever the price of that bundle differs from the price of the associated derivative.

This latter opportunity is facilitated by the second major change we have seen in markets, namely, the ability to get information about the trading conditions across a vast array of securities and markets, calculate potential opportunities, and execute any needed trades, all in the blink of an eye. This revolution in trading is due to the rise of the computer and, indeed, things now happen far faster than the blink of an eye (which takes a somewhat poky 350 milliseconds, a length of time in which an electron can travel more than 65,000 miles).

The combination of highly connected markets and quickly conducted trades has formed a new kind of complex system unforeseen even a decade ago. A trade in one market reverberates across the others, as the various inconsistencies it induces get corrected. Of course, those corrections can start their own reverberations. If the (unintended) feedback loops emerging from the various connections are negative, the reverberations in the system slowly die out, and the markets are better for the experience, as prices realign with one another. If, however, the feedback loops are positive, we end up with the reverberations amplifying one another, creating something akin to the horrible screeching sound we hear when a microphone is held too close to a loudspeaker.

Over the past few years, a new kind of trading firm has arisen on the market scene: the high-frequency trader (HFT). These firms have fully embraced the information age and have created algorithmic traders that watch over markets and execute any desirable trades on a remarkably short time scale. For this type of trading, getting your buy or sell message to the exchange before anyone else is so important that factors such as where you physically locate your computer hardware matter—an electron travels about a foot each nanosecond, so every foot closer to the exchange's machines gives you a nanosecond advantage over your competitors.

HFTs now account for an enormous volume of trades. In general, they don't like holding too many shares at any one time. Thus, while they might buy a lot, they also sell a lot, so at the end of the day (though the new reality is that with global, interconnected markets, there really is no end of the day) their net holdings of any given security are small.

The existence of HFTs certainly alters the dynamics of markets. In the late 1980s at the Santa Fe Institute, my colleagues Richard Palmer, John Rust, and I created the Double Auction Tournament to test some core ideas about markets. Academics, professional traders, and interested amateurs tested their trading strategies over the web (to our knowledge, this was the first Internet-based auction) and then submitted final versions to us for analysis.

One of our interests was what would happen in a hybrid market composed of both machine and human traders. When we ran such a market without any compensation for the innate speed differences between humans and machines, we found that when the market opened there was a flurry of trades by the machines before the humans could even react. After that point, the machines just stayed quietly in the background as the humans traded with each other, activating only when a human made a bad offer, in which case a machine would jump into the market and steal the deal.

One suspects that our current market system, with both human traders and HFTs, may behave in an analogous way to the hybrid Double Auction Tournament. HFTs' distinct speed advantages may be causing flurries of machine trading, punctuated by quieter periods where the machines remain in the background waiting to take advantage of human errors. When we eliminated the speed advantages of the machines, humans and machines easily coexisted and were difficult to tell apart given the data, with the one exception being that humans tended to place offers that ended in digits of either zero or five, while machines were not so constrained.

Returning to that fateful day in May, at 2:32 p.m. our trader presumably pushed the enter key in response to some innocuous-sounding prompt along the lines of "Are you sure you want to execute these trades?" and a stone was dropped into the market pond, causing a small ripple. The market was able to absorb the initial trade volume, as HFTs and other intermediaries bought the newly offered contracts. Over the next ten minutes, the HFTs accumulated quite a few of these contracts, and in order to balance out their positions, they began to sell. A game of high-stakes hot potato ensued, in which the HFTs began to buy and sell to one another, with only occasional leaks of the contracts out to other market participants.

It is at this point that the fatal flaw in the algorithm becomes apparent. The game of hot potato started to generate a lot of market volume, with more than 100,000 shares being exchanged in a very short time. The algorithm, blind to everything but the volume, saw this increased activity as a sign that the market was liquid and that prices were stable, and it began to dump even more shares into an already volatile mix. This new action destabilized things even further, as any real liquidity in the E-mini shares dried up and the prices began to plummet. In the thirteen minutes after the enter key was pressed, the algorithm sold 35,000 contracts, and the remaining 40,000 contracts were sold off in a scant seven minutes more. Thus, all of the initial 75,000 contracts were sold in under twenty minutes—whereas in the past, using more standard algorithms, it had taken around six hours to dispose of similarly sized lots. The initial trades and subsequent activity, not surprisingly, resulted in a substantial drop in the price of the E-mini contracts (see Figure 3.2).

However, at 2:45:28 p.m. an event happened that likely prevented an even deeper disaster. At that time, an automated mechanism paused trading for five seconds. This mechanism had been put in place by the exchange, and it was designed to recognize market conditions in which the execution of further trades would result in unnaturally large price swings. While five seconds seems like an inconsequential amount of time, it is an eternity in an era when nanoseconds rule. It was long enough to allow other traders to enter the market and get things on the road to

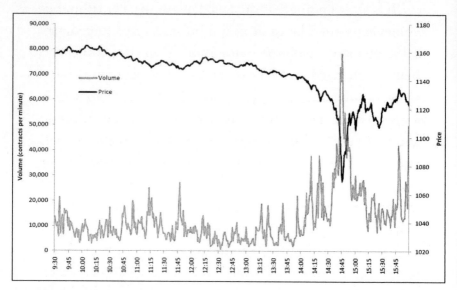

FIGURE 3.2: The price and volume of E-mini contracts on May 6, 2010. *(Source: US Securities and Exchange Commission.)*

recovery. Over the next twenty-three minutes, buyers with a more fundamental focus began to flood the market, and prices rebounded.

The proximate cause of the bad behavior in the E-mini market is easily tied to the flaw in the trading algorithm. By linking the number of trades to only volume, a positive feedback loop was unintentionally embedded in the algorithm: if the initial trades cause a big increase in volume, then the algorithm trades even more, which will further increase volume. If the HFTs had not been in the picture, the naive trading algorithm might not have induced enough extra trades to trigger the feedback loop. However, with the rapidly trading HFTs and their desire to maintain relatively neutral share positions, a new market dynamic formed that embedded a positive feedback loop into the system.

If this were just the story of the E-mini market, it would be worth telling as a parable about algorithmic—actually, human—hubris and the dangers of unintended consequences and positive feedback. But the story does not end here.

Given the interconnectedness of markets, what happens in the E-mini market does not stay there. As the E-mini declined in value, traders started to look for arbitrage opportunities elsewhere—in this case, either in SPDRs or in the stocks that make up the index itself. While the E-mini's price was rapidly declining, driven by the positive feedback loop, the prices of SPDRs and the stocks that formed the index moved much more slowly. This created a new opportunity to profit by buying the relatively cheap E-minis and selling their more expensive equivalents in the form of SPDRs or the bundle of underlying stocks.

In a well-functioning market system, the arbitrage opportunities created by the collapsing E-minis would normally dampen the price dynamics. The profit-seeking activities of the arbitrageurs would raise the price of the E-mini (given the newfound demand to buy) and lower the price of the SPDR or bundle (given the newfound desire to sell), and the prices would realign and remove the opportunity for profitable arbitrage. Unfortunately, given the preexisting turmoil and the positive feedback loop, the markets failed to realign very quickly, and the arbitrage opportunity remained. This resulted in trading pressure on the other markets, and they started to eat into their respective order books as well. Moreover, the newly generated chaos made many potential market makers nervous, as nothing in the incoming data streams—which by this time were starting to falter given the massive influx of trading—could account for the large price changes being observed. This triggered data integrity checks, where firms paused their trading activity. Other firms withdrew entirely from the market as automatic systems that continuously monitor a firm's position and potential exposure to financial risk began to exceed preset limits and halted the firm's trading. Finally, in some firms, humans overseeing all of this bizarre activity simply lost their nerve (or behaved wisely) and withdrew their offers from the market.

As the market makers withdrew, the order books began to empty out, leaving only long-standing orders and, at the extremes, automated "stub" orders set at ridiculous price points just to ensure that there would always be someone willing to buy or sell any given share. Thus, the transactions that did occur were happening at prices that became more and more extreme over time. More than three hundred stocks experienced price changes of as much as 60 percent (more than 20,000 trades, constituting 5.5 million shares, were executed at such extremes). At the most extreme, securities were traded at their stub prices, with some shares going for a penny and others for $100,000.

The aftermath of the events of May 6, 2010, was significant. In the short term, there was a realization that the events were far from the "fair and orderly" markets that the exchanges want to oversee, and the trades that took place far from the prices prevailing just before the chaos began were broken by the exchanges, as they were considered "clearly unrealistic prices" that were "clearly erroneous" given the severe market conditions. While exchanges have always had the power to break such trades (always read the fine print), the actual mechanisms used for determining "clearly erroneous" were not well defined, and this has prompted a reform in this area. The second major reform has altered how various circuit breakers get deployed. Individual markets often have mechanisms designed to halt trading when unexpected conditions arise, and in practice, even very short halts have allowed markets to stabilize quickly and resume in an orderly fashion. Unfortunately, even the existence of circuit breakers can have unintended consequences, as multiple halts in a given security might cause market makers to withdraw their liquidity. Also, given global connectivity and many markets trading the same security, a trading halt in one market might just shift the displaced trades to another market, circumventing the original breaker and exacerbating the problem.

The one area that has not been reformed is limiting the HFTs. For example, the feedback loops induced by the

HFTs could be dampened by imposing transaction taxes or redesigning markets to lessen the importance of nano-second-scale speed.

Even the above repairs do not address the fundamental problem that caused the flash crash. We have unknowingly created a complex adaptive financial system that we do not understand and cannot control. At each stage of its creation, we have accrued additional complexity in the name of added benefits: connecting markets with one another will ensure that price discrepancies will be eliminated quickly, having high-frequency traders will guarantee a ready trading partner for any transaction, using derivatives will provide a means for farmers to hedge the risks of bad weather and for pension funds to insure their portfolios, and so on. While each of these individual pieces makes sense, the collection may not.

As we have already seen, reductionism does not imply constructionism. Thus, while the motivation for, and understanding of, any single piece in the system may be sound, that should not give us any confidence in the behavior of the whole. The flash crash occurred not by design but through emergence.

The flash crash was a surprisingly gentle warning that we must heed. The events during that thirty-minute period in May, while striking, were reversible. While careful autopsies of dramatic events are useful, we need to be in a position to prevent the appearance of the bodies in the first place. Unfortunately, the flash crash has shown us that, however good our retrospective investigations might be, our prospective knowledge is weak. We can't even begin to grasp the implications of the financial systems we have built.

While the flash crash was driven by greed in the pursuit of profit, it fortunately involved ignorance, not malice. Imagine the chaos and long-term devastation that could happen if malice and a bit more forethought were involved. How difficult would it be for, say, a terrorist organization or rogue state to infiltrate either the computer or human systems that underlie our markets and wreak havoc on a much larger, and longer-lasting, scale? This does not seem all that hard. Attacks on the cyber infrastructure, such as cracking the actual systems of the exchange or those of the numerous decentralized trading operations, or somehow disrupting or altering the communication flows that direct or report trades, seem possible, especially given examples such as the Stuxnet computer worm, which hampered Iran's ability to enrich uranium. The human systems connected with financial institutions are also vulnerable. Indeed, there are examples where the actions of a single trader brought down an entire institution, as with the fall of the 233-year-old Barings Bank in 1995. Thus, inserting one or more traders into the system with enough access to the trading desks to launch a carefully coordinated, malicious attack is feasible. A more ambitious approach might include setting up an apparently legitimate fund or HFT operation that gets privileged and unfettered access to the trading systems, or, if that is too bothersome, simply executing a large number of simultaneous transactions spread across legitimate traders. The impact of such an attack is hard to predict, but at the very least it would seriously erode confidence, and it could be far more consequential, leading to a partial collapse of the very markets that ensure our economic survival.

Unfortunately, the story encapsulated in the flash crash may not be all that unique. Indeed, the recent worldwide financial collapse that started in 2008 has similar undertones.

At the heart of the 2008 financial collapse was an economic crisis that fully embraced all of the seven deadly sins. Gluttonous fixed-income-asset buyers, for the promise of slightly higher returns, were willing to buy up newly formed collateralized debt obligations. Extravagant home buyers, hoping that rising house prices would allow refinancing in the future, opted for houses and ballooning mortgage payments well beyond their current means. Greedy mortgage brokers, able to pass on even suspect mortgages to firms that created and quickly sold off mortgage-backed securities, were willing to qualify almost any buyer. Envious firms, wanting to boost their bottom lines, began leveraging themselves while marketing suspect derivatives to their customers. Slothful rating agencies, relying on the word of the firms and outdated statistical models, gave absurdly high ratings to novel securities while collecting commissions. Prideful government agencies, relishing the increase in home ownership and the power of the unregulated market, stood idly by. As for wrath, hell hath no fury like a complex economic system scorned.

The point of the previous paragraph is not to tell some modern morality tale but rather to emphasize how, at each level of the system, the entities involved were following perfectly understandable—though perhaps not virtuous—incentives. Thus, in a very real sense, economists and policy makers were fully equipped to understand each *part* of the system. Unfortunately, as we have seen before, thinking that understanding the parts of a system implies that you understand the whole system is a sin that is committed all too often.

As we saw in the case of the flash crash, positive feedback mechanisms amplify small events into large ones. The housing market is rife with positive feedbacks. If mortgage money becomes easier to get, the demand for houses goes up, resulting in higher house prices. These higher house prices make lenders more willing to grant mortgages, as rising prices ensure that sufficient collateral exists to lower the risk of the loans.

In the US housing market, the positive feedbacks tended to reinforce every part of the system. Higher house prices encouraged more buyers, lowered lending standards, and resulted in less risky derivatives and easier government policies, and each of these fed back on the others, reinforcing the chain of effects. Alas, the same forces that amplified the system on the way up accelerated its demise on the way down. Unfortunately, unlike with the flash crash, there were few circuit breakers, or anything like them, in place during the financial collapse.

The interactions and connections among the various parts of the system are critical here. Imagine any of the key markets associated with the financial collapse as a timber farm along a lightning-prone ridge. Every now and then, lightning strikes, and if it hits a tree, that tree goes up in flames and ignites any neighboring trees. If you want to maximize the timber harvest, you must make a trade-off between growing more trees to get more timber and keeping land fallow to contain neighboring fires. The best choice here depends on various underlying factors, such as the frequency of lightning and the growth rate of trees, but whether the best choice gets made depends on who owns the ridge. If a single person owns the ridge, it will be in her interest to include a few firebreaks, so that a single spark won't lead to a conflagration that takes out the entire ridge. Unfortunately, such firebreaks may not arise in a system where each potential tree site is owned by a different individual following her own incentives. In this situation, while all individuals would benefit from the inclusion of firebreaks, no individual wants to be the person who provides the firebreak, since she would have no timber to harvest. In economic terms, firebreaks are underprovided, and this results in far more destructive fires and much lower harvests than are possible under a more coordinated regime.

So it was with mortgages at the start of the financial crisis. No entity wanted to forgo any possible trade and lose some immediate profit. Thus, a single bank finds it individually profitable to hold securities issued by another bank, even though that other bank has bought securities from another bank, and so on down the line, to the point where the failure of a very distant bank can cause the whole system of promises to unravel. Similarly, a single firm may simultaneously buy and sell insurance-like policies on the risk of default (known as credit default swaps) and feel that its position is safe, since any loss to one of the policies will be perfectly offset by a gain to the other. However, if one firm fails to meet its obligation to pay in the case of a default (think American International Group, aka AIG), the entire system unravels. In these and countless other situations, what is important here is the chain of connections that results from individually rational, but globally irrational, arrangements. Without well-placed firebreaks, these systems are subject to small events having catastrophic consequences.

In both the 2008 financial collapse and the flash crash, we saw systems that were vital and thriving at one moment suddenly become quiescent. This type of switching happens in a variety of complex systems. For example, a living organism exists in a dynamic state where its many interacting parts result in a vital and robust organism. Now introduce, say, a well-placed shock, and the once vital organism is pushed into a death state where none of its parts interact. Unfortunately, this too is a robust state.

Expectations often keep social systems, and especially markets, operating. Expectations can lead to self-fulfilling prophecies, both good and bad. Thus, in a flash crash, once liquidity dries up, the expectations of market makers may change to the point where they believe they will no longer be able to find reasonable trading partners, which causes them to withdraw their orders and realize their expectations, further exacerbating the liquidity crisis. Once a housing bubble begins to pop, the downward spiral of house prices alters the expectations of the lenders, and they become wary of granting new (or refinancing old) mortgages without extreme levels of collateral, which in turn causes prices to fall and the newly formed expectations to be reinforced. In both cases, feedback loops concerning expectations exacerbate a bad situation.

When emergence is working for you, the invisible hand of Adam Smith is a wonderful thing. Life would be a lot more fun, albeit much less intriguing, if emergence arose only when it led to good things. Unfortunately, we have seen the dark side of emergence, in which a seemingly innocuous event triggers a cascade that leads to disaster. Complex systems, whether intentional or not, are playing an increasingly important role in our world. While we might not ever be able to fully control such systems, we may be able to mitigate their downsides through the clever introduction of metaphorical firebreaks such as the circuit breakers that are used in financial markets. Our understanding of how to create such controls is lagging well behind our need to implement them, and we must quickly develop this knowledge so that the kingdom won't be lost for want of a nail.

FOUR

From One to Many: *Heterogeneity*

> One Ring to rule them all, One Ring to find them,
>
> One Ring to bring them all and in the darkness bind them.
>
> —J. R. R. TOLKIEN, *The Fellowship of the Ring*

ECONOMISTS ARE FOND OF THE "REPRESENTATIVE AGENT," A theoretical convention that makes the math much, much easier. The idea behind the representative agent is that instead of having to worry about, say, every consumer in the economy, we can substitute a single consumer to represent everyone—one agent to rule them all, as it were. Obviously, such an assumption greatly simplifies the resulting model, as the representative agent can stand in for a vast horde of individually quirky consumers who might be difficult to track one by one. Indeed, theoretical economists and policy makers often use such a trick in models that influence the lives of hundreds of millions. As long as individual behaviors average out appropriately, using such an approach seems like an obvious choice.

Whether we can use representative agents to model complex systems is really a question about whether heterogeneity matters. If it doesn't matter, then assuming average behavior embodied in the form of a representative agent suffices: the same behavior will emerge from a system modeled by a population of actors as from one consisting of a single representative agent. If it does matter, then we need a new approach to understand, predict, and control our world.

Jane Jacobs, in her remarkable book *Cities and the Wealth of Nations*, admonishes economists to get the answer right here:

> We think of the experiments of particle physicists and space explorers as being extraordinarily expensive, and so they are. But the costs are as nothing compared with the incomprehensibly huge resources that banks, industries, governments and international institutions like the World Bank, the International Monetary Fund and the United Nations have poured into tests of macroeconomic theory. Never has a science, or supposed

science, been so generously indulged. And never have experiments left in their wake more wreckage, unpleasant surprises, blasted hopes and confusion, to the point that the question seriously arises whether the wreckage is reparable; if it is, certainly not with more of the same. Failures can help set us straight if we attend to what they tell us about realities. But observation of realities has never, to put it mildly, been one of the strengths of economic development theory.

Consider a honeybee hive. Every egg laid by the queen goes through a delicate sequence of development from egg to larva to pupa to finally emerging from its honeycomb cell as a fully formed bee. For this sequence to be successful, it requires a narrow range of temperatures to be maintained inside the hive (close to 94 degrees Fahrenheit). Of course, the temperature outside the hive varies wildly, so how can bees keep the inside temperature confined to such a small range?

It turns out that worker bees have two temperature-related behaviors. When a worker gets too cold, it seeks out other bees and rapidly buzzes its wings to generate heat. When it gets too warm, it moves away from others and fans its wings to form air currents that will cool things down (see Figure 4.1).

The temperature in the hive depends on the actions of its workers. There is no central command center in the

FIGURE 4.1: Worker honeybees spread out at the entrance of their hive and fan their wings to create air currents that will cool the hive. This behavior is activated by a genetically determined set point. *(Photograph courtesy of Jacob Peters.)*

hive, and it is only through the decisions and actions of each individual bee that things get done. It turns out that an individual honeybee's temperature-related behavior is given by a genetically determined set point. Temperatures much above or below this set point cause the bee to undertake cooling or warming behavior, respectively.

Temperature control seems like a situation in which a population would benefit from homogeneity—in which nature would evolve a representative agent. Researchers at the University of Sydney (see Jones et al., "Honey Bee Nest Thermoregulation: Diversity Promotes Stability," *Science*, 2004) investigated this question and found a surprising result.

As a thought experiment, suppose we observe a hive of bees in which every bee's genetic thermostat is set to the same ideal temperature. You might think that since all of the bees are so precisely calibrated, the hive will maintain a constant temperature. This is not what happens. When the temperature creeps below the set point, large numbers of bees instantly huddle together and buzz their wings, causing a large increase in the temperature. As the temperature rises, it quickly goes past the ideal point, and all of the bees switch to their cooling behavior and scatter and fan, inducing a rapid drop in temperature. As the temperature plummets below the ideal point, the mass of bees switches behavior yet again. What emerges is not a hive with a tightly controlled temperature but one that experiences wild swings in temperature.

As an alternative, suppose that we have a hive of heterogeneous bees, each with a slightly different set point around the ideal temperature. In this hive, as the temperature starts to creep below the ideal point, only a few bees start to huddle together and provide a little additional warmth, slowly raising the temperature. Indeed, any time the temperature overshoots or undershoots the ideal point, there is a graduated response by the bees, with only a few joining in at first, and more joining in only if things start to stray further from the ideal. Ultimately, this heterogeneous strategy allows the hive to maintain a precise temperature with only minimal oscillations.

Thus, having a heterogeneous population of honeybees is adaptive to the hive, leading to a much more tightly controlled temperature and greater success in brood rearing. In real hives the virgin queen spends her first few days going out on flights where she mates with around eight to twenty drone (male) honeybees from different hives, rather than just one. Once the queen is back in her hive, she produces worker bees that are either sisters or half-sisters to one another, guaranteeing some heterogeneity among them.

The *average* temperature set point of the honeybees was the same in both our homogeneous and heterogeneous hives. The difference was that in the heterogeneous hive there was some variance of the set points around this average, whereas in the homogeneous hive every worker had the same set point. So, at least in terms of a honeybee hive, the representative agent model would be very misleading, implying hive temperatures that oscillate wildly when in fact they are actually quite stable.

Now consider a model of a market. Let's assume that the market is populated by homogeneous representative traders who decide to buy or sell based on incoming information. Just as we saw with the honeybees, this type of model is going to result in some unusual market behavior. As the information in the market begins to change, at some point the representative trader is going to want to buy. Since all of the traders use the same rule, this is going to cause a drastic increase in demand, and prices will experience a rapid rise. As prices go up, the information eventually changes to a point where, in perfect synchronicity, all of the traders want to sell, inducing a price crash. As in the case of the hive, a market with homogeneous traders leads to wild price oscillations.

Stable markets emerge only with heterogeneous agents. With many types of traders, responses to changing information are graduated, with slight changes in information influencing only the most sensitive traders, and more extreme changes provoking responses from the less sensitive traders. Such a market will be much better behaved than a homogeneous one, experiencing milder price swings and more reasonable "price discovery."

In hives and markets heterogeneity provides needed stability, but this is not always the case in other systems. Suppose we want to model the dynamics of a social movement, ranging from a neighborhood-level riot to the overthrow of a national government. Let's assume that each of, say, one hundred people in our society has a sensitivity level, S, such that if she observes S or more people participating in the movement, then she will join. Finally, let's assume that there is a group of outside rabble-rousers that tries to start the movement.

Assume that our one hundred people all have the same sensitivity level set at, say, 50. How many rabble-rousers will it take to trigger an all-out social movement? If the number of rabble-rousers is less than fifty, then no one else joins in the fray. If the number of rabble-rousers is fifty or above, then everyone joins. Thus, in a homogeneous world, it takes at least as many rabble-rousers as the fixed sensitivity level to catalyze a movement. In this example, we need a fairly large number of rabble-rousers—equal to half of the population—before we see a full-blown social movement.

Alternatively, assume that we have a very heterogeneous population, with each of our one hundred people having a unique sensitivity. To make this an extreme example, line up the population and give the first person a sensitivity of 1, the second a sensitivity of 2, and so on down the line, until the last person is assigned a sensitivity of 100. In this world, how many rabble-rousers are needed to catalyze a society-wide social movement? The answer, of course, is one. One single rabble-rouser is enough to get the person with a sensitivity of 1 to join in, and once we have two people in the movement, that is enough to get the person with a sensitivity of 2 to join, and this triggers the third (which, according to Arlo Guthrie's song "Alice's Restaurant," constitutes an organization), and so on down the line, until all one hundred members of our society have joined the movement.

Both of the social worlds above are characterized by a critical tipping point, whereby below this point no one joins the movement and above it everyone does. Of course, this tipping point is dramatically different in the two worlds, being equal to fifty (half the population) in the first and only one in the second. Note that in both worlds, the *average* threshold for the population is about fifty, so the different tipping points are due to the variations in the thresholds of the two worlds. In the first world, the presence of homogeneous agents implies no variance, while in the second agent heterogeneity induces a lot of variance.

Thus, in the social movement model, we find a case where heterogeneity leads to instability rather than stability. However, both the bees and the protest share important characteristics that underlie the dramatic difference in outcomes. In both cases, heterogeneity leads to a graduated response, where slight changes in the environment cause slight changes in the system's behavior. The difference between the models is in the type of feedback they engender. In the case of hive temperature regulation, the system is governed by negative feedback, and having a graduated response tends to stabilize the system. In the case of the social movement, there is positive feedback, and a graduated response is like a rolling snowball, where the accumulation of snow makes it bigger and heavier and more likely to pick up additional snow.

Despite the type of feedback in play, both models make the same essential point about representative agents: they can be quite misleading, as the mean is not the message. If we consider systems with all agents acting at the mean, we will often make bad predictions, expecting too little stability in the case of the honeybees and too much in the case of social movements.

Policy can often influence the level of heterogeneity in the system and thus determine the system's overall behavior. Heterogeneity is likely to be a stabilizing force in markets, and therefore we might want to encourage diversity by ensuring that we have many moderately sized trading houses competing with one another using proprietary trading algorithms. However, if you want to quash a social rebellion, having a homogeneous population with a high threshold will prevent small events from growing into revolutions. While policy can't dictate a homogeneous population, it can influence the feedback loops by, say, altering the information individuals receive about reasonable threshold levels or the number of activists. Alternatively, if you want to initiate a social movement from a small spark, then you want to encourage a diversity of views and a sense that everyone is participating, so that a single spark can lead to a cascade that ignites a full-blown movement—it takes just one ring to bring them all.

FIVE

From Six Sigma to Novel Cocktails: *Noise*

> Errors . . . are the portals of discovery.
>
> —JAMES JOYCE

SIX SIGMA IS A BUSINESS MANAGEMENT SYSTEM DEVELOPED BY Motorola starting in the 1980s. It was designed to improve manufacturing processes. At its heart is a set of techniques that attempts to limit the number of defects in a process to 3.4 per million or fewer (or, equivalently, having 99.99966 percent of the products emerging from the process defect free). The application of Six Sigma ideas and, more generally, the notion of improving quality by eliminating errors have likely resulted in substantial savings to producers and increased benefits to consumers in industries ranging from microchip fabrication to health care.

Given examples such as Six Sigma and our own intu-itions, it is easy to think that eliminating errors in a system—what often gets classified as "noise"—will lead to a better outcome. Manufacturing is in many respects the opposite of an emergent system. It is a system that thrives on homoge-neity. But, as we saw in the previous chapter, there are times when you need heterogeneity to drive a system. While er-ror avoidance is useful in the manufacture of a well-defined good, it is a dangerous bias if we want to discover new things.

Consider the problem of finding the highest elevation on some landscape. As you tread across the landscape, you change your latitude and longitude with each step, and if the landscape has hills and dales, you'll also change your elevation. Such elevation seeking is an example of a simple search problem (searching for the highest elevation) across two dimensions (latitude and longitude).

If the day is clear and we can take a hot-air balloon journey and rise above the landscape, or if we can quickly scan a topographic map (composed of contour lines of el-evation spreading out like ripples around various hills and dales), then finding the highest elevation is fairly easy. With either type of view we can quickly identify the highest point in the land and find its associated latitude and lon-gitude. In such a situation, with either view of the world, the error-eliminating Six Sigma approach would work like a charm.

To make this scenario a bit more challenging, suppose a dense fog rolls in, thus limiting our vision to just a few feet surrounding our current position. Under such conditions—which are the norm in real-world search problems—what can we do?

An obvious search strategy here is to simply look around our current fog-bound location and take a step uphill. Once we take that step, we can look around anew, as a bit of new territory will be revealed, and we can again step uphill. At times we might look around and find that everything is the same elevation, and if so, we can just step in a random direction. As we continue to follow this search strategy, we will eventually find ourselves at a point where, as we look around in the dense fog, all directions lead downhill. Here we note our coordinates and declare to the world that we have found a high point. This type of search strategy is known, not surprisingly, as hill climbing.

How well does hill climbing work? If, when we look around, all roads lead downhill, we can at least guarantee that when the fog lifts we will be at a *local* high point. There is no guarantee, however, that this local high point will also be the global high point. Thus, while we might declare at the end of our fog-bound climbing efforts that we have found the top of the world, when the fog lifts we might find ourselves standing on an anthill at the base of Mount Everest.

The problem with a hill-climbing search is that we might end up at a local, rather than global, optimum at the end of our journey. One way to improve our odds of finding the higher points on the landscape is to do multiple hill climbs, each starting from a different, randomly chosen location. Return to our fog-encrusted landscape and think about randomly parachuting down a few hill climbers, each of whom pursues a hill-climbing algorithm from wherever he or she lands. If there are many hills on the landscape, then these different climbers are likely to end up atop different peaks, some higher than others. Thus with multiple, random starting points we are likely to uncover new and better optima.

The effectiveness of hill climbing is tied to the ruggedness of the landscape. If the landscape looks like that around Mount Fuji, dominated by its symmetrical volcanic cone, then regardless of where a hill climber lands, when she walks uphill she will end up at the peak and find herself at the highest point. If, instead, the landscape resembles that of the Himalayas, Andes, or Rockies, then it will be quite likely that our hill climbers will end up on local rather than global peaks.

Consider the one-dimensional problem shown in Figure 5.1. For any location along the x-axis, there is an associated elevation given by the y-axis. Note that we can take any point on the x-axis and figure out where a hill climber who starts from this point will end up after marching uphill. Alternatively, we can take any peak on the landscape and map out all of the x-axis values that will lead, after hill climbing, to it. Such a map gives us the "basin of attraction" for each local optimum. If the world is like Mount Fuji (think of a single pyramidal shape dominating the diagram), then all of

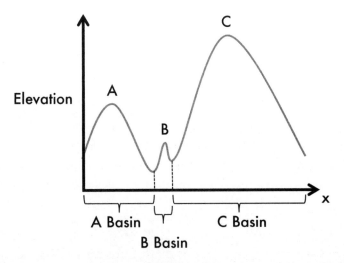

FIGURE 5.1: A one-dimensional search problem, along with its associated basins of attraction (assuming hill climbing). As the landscape becomes more "rugged," the number of peaks, and thus the number of basins of attraction, increases.

the x-axis values are in the same basin of attraction, and that basin leads to the global high point. If, instead, the world looks like that of the figure, then there are three basins, each leading to a different peak with a different height.

One can also identify basins of attraction in landscapes with more than one dimension. In the case of the two-dimensional landscape we initially discussed, think about a great flood inundating the world. As the waters begin to evaporate, the first bit of land that emerges will be the global peak. As the waters move down the sides of this peak, the associated basin of attraction is revealed. If the landscape is rugged, at some point another island of land will emerge from the receding waters, revealing the second-highest peak, and we can begin to identify its basin of attraction. As the waters fall further, the various islands of land that are emerging will eventually touch one another, and it is at these points that the boundaries of the basins of attraction are identified. If, as the waters recede, we have a single point of land growing to encompass the entire world, then hill climbing will easily find the global peak. If, instead, numerous islands pop up, then hill climbing will likely get trapped on a lower-lying peak.

From the above, we see that the ruggedness of the landscape is an important factor in the ability of hill climbing to discover the highest point. This leaves open two key questions: what determines ruggedness, and are there better ways to search than hill climbing when we confront a rugged landscape?

The issue of ruggedness is tied to the predictability of elevation changes as we traverse the landscape in any given direction. If we start at a random location on the edge of the landscape, pick a random direction, and start walking in a straight line, we can keep track of the number of times we switch from ascending to descending, and vice versa. Landscapes will be relatively smooth if these traverses tend to have very few switches, and rugged when they have a lot. When we have very few switches, the coordinates (dimensions) of the search are relatively independent of one another. That is, if along our traverse the gains or declines in elevation are not tied to where we are on the landscape, then the landscape will not be rugged. However, if the dimensions of the search start to react with one another—that is, if the elevation change that I experience when I change my longitude by a little is closely tied to my current latitude—then the landscape will be rugged.

Systems that have a lot of interaction among their various dimensions are known as nonlinear systems. Oddly, there is a specialized area of science devoted to the study of nonlinear systems. The reason this is odd is that some degree of nonlinearity is present in almost all real-world systems, so science treats as some sideshow curiosity an aspect of the world that is actually the norm. This observation was captured in a remark apparently made by the mathematician Stan Ulam: "Using a term like nonlinear science is like referring to the bulk of zoology as the study of non-elephant animals."

The notion of interacting dimensions and ruggedness gets far more interesting if we consider search problems other than finding the highest point on a physical landscape. For example, think about trying to dress in a fashionable way. Some of the potential dimensions here might be the style of the outfit, its color, choice of belt, type of shoe, and so on. If these dimensions don't interact, then leaving the house in a fashionable outfit is relatively easy. First you find the best belt among the lot. Then you pick the most fashionable shoes. Next, pick the best color. And on and on, until you leave the house dressed in the ideal combination.

Of course, in fashion, as in life, different dimensions do interact quite a bit. The choice of shoes depends on the color and type of outfit, the belt needs to be coordinated with the shoes (or so I'm often told), and so on. Thus, optimizing on each dimension alone and ignoring the others is likely to lead to an overall ensemble that is a fashion faux pas. Moreover, there are likely to be many different ensembles that work well together, each representing a local optimum (style) that cannot be improved upon by making minor changes in any one element. Perhaps one of these ensembles is better than all of the others, and we can find the ultimate in fashion-forward thinking, though more likely than not these radically different ensembles may each solve the problem of looking fashionable with roughly equal success.

A lot of other choices tend to be characterized by rugged landscapes. Consider finding the best design for a car. The various features of a car, such as the number of doors, presence or absence of tail fins, size of the engine, hard or rag top, wheelbase, weight, mileage, and so on, all interact with one another in surprising ways, and thus the space of car designs is likely to be very rugged. Out of such a landscape, a Ferrari 250 GTO, a Toyota Corolla, and a Ford F150 might all emerge as locally optimal solutions to the problem of car design. Similarly, consider the problem of finding the best cocktail, in either the mixed-drink or drug sense. If the various elements of the cocktail do not interact, then we can just optimize on one element at a time, find its ideal, and from such a search combine all of the ideal points to make the ultimate cocktail. Of course, such an approach tends to result in cocktails that are fairly unappetizing in the case of mixed drinks (with perhaps a rare counterexample given by the Manhattan iced tea) and ineffective and (likely) dangerous in the case of drugs.

While hill climbing could lead us to a Jeep and a pair of hiking boots to explore Canyonlands National Park and a Bentley and a tuxedo for a formal, it has the potential to leave us trapped on much less desirable peaks. One escape from the traps of hill climbing on a rugged landscape is to introduce noise or error into the search algorithm. Our intrepid hill climbers above followed the Six Sigma mantra and always headed uphill without the possibility of error. However, what's good for refining a manufacturing process is likely bad for discovery.

Return to our fog-bound, elevation-seeking hiker who is standing on an anthill at the base of Mount Everest. When she looks around from her current perch, she sees only opportunities to descend. If she is to escape the mound and climb Everest, she must take some downhill steps. This implies that she will need to make an error in her hill-climbing search and take at least one step downhill. While this is seemingly at odds with her overall goal of finding the highest point possible, this short-term loss offers the long-term potential of moving her to a new slope that might result in her discovering a much higher peak.

A search algorithm that embraces this hill-climbing-with-error idea is known as simulated annealing. In real annealing, the properties of a material such as glass or metal are improved by heating the object and then allowing it to cool slowly. Within such materials, the individual atoms have a tendency to align with one another, all else being equal. When heated, this tendency gets overwhelmed by the noise introduced by the outside energy, and the atoms flop about in all directions. If the heated material is quickly cooled, the atoms get frozen in whatever scrambled directions they were facing at the time of the quenching. However, if the material is cooled very slowly, as the flopping about of the atoms slowly diminishes with the lowering temperature, their desire to align with one another begins to take over. Eventually the material cools to a state in which most of the atoms are aligned. Such crystalline structures often give the resulting material desirable properties.

Simulated annealing uses an idea very similar to that of real annealing. We take our standard hill-climbing algorithm, with its desirable tendency to march uphill, and impose on it some noise via a high "temperature." While the algorithm still wants to march uphill, the noise makes it willing to accept occasional errors and march downhill as long as the temperature is high or the loss of elevation is small. Over time, we reduce the temperature, lessening the algorithm's tendency to take large downhill steps, until the temperature is so low that the algorithm reverts to its pure hill-climbing behavior.

The intentional introduction of such errors into the search process gives us the ability to overcome the usual traps of rugged landscapes (see Figure 5.2). In essence, the noise we introduce begins to vibrate the landscape, filling in the small valleys enough so that the searcher can traverse these previously insurmountable obstacles on her way to higher ground. Noise allows the hiker to step off the anthill and proceed up Everest. There is, as in most things, a trade-off here. The addition of noise, at least in the short term, tends to lessen performance, and on easily surmounted landscapes such as Mount Fuji, it adds additional time to the search, as well-directed uphill steps are occasionally counteracted by downward ones. Of course, the benefit of noise is that on more rugged landscapes it

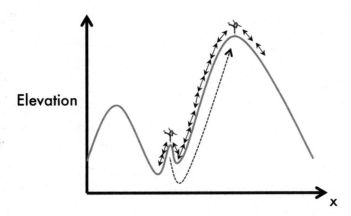

FIGURE 5.2: The occasional downhill steps introduced by simulated annealing allow the hiker to escape inferior local optima and ultimately discover the global optimum.

results in much better performance overall, as it allows the climbers to escape low-lying local optima.

A nice application of the above ideas is to the problem of discovering novel and effective cocktails of drugs for chemotherapy. Drug discovery has taken many forms over the years. Sometimes drugs were isolated from folk remedies, such as the bark of the willow tree, which gave us aspirin. At other times we found new drugs by collecting chemical compounds from various plants, animals, and microbes and then screening these compounds against various diseases in the hope of finding something that works, such as the fungus that resulted in penicillin. More recently we have tried to design drugs directly, by first identifying a potential molecular weakness in a disease and then designing an appropriate drug to attack this vulnerability—this typically comes down to identifying the shape of a protein needed to exploit the weakness, and then generating that shape using knowledge about how various combinations of molecules fold into three-dimensional objects.

All of these approaches have resulted in useful, albeit at times expensive, drugs. Unfortunately, the hope of finding a *single* drug to cure each disease assumes that the disease has an Achilles' heel that can be attacked by the chemical equivalent of the arrow of Paris. However, most diseases exist in the realm of complex biological systems, and such systems tend to have built-in redundancies that make the system as a whole robust in the face of a single line of attack. While such systems will not succumb to a single attack, they are vulnerable to a salvo of arrows each destined for a different target, with the overall goal of simultaneously neutralizing enough of the redundancies so that the system as a whole will fail. This complex-systems view of disease suggests that drug cocktails, with each drug acting like an arrow in the aforementioned salvo, may be needed if we are to cure what ails us.

Perhaps the best-known drug cocktail is the set of antiretroviral drugs used in the treatment of HIV/AIDS. The human immunodeficiency virus undergoes rapid mutation, and so while targeting only one part of the virus will provide some short-term success, this strategy will ultimately fail, as mutations will arise that circumvent the initial target. However, by attacking the virus with several drugs at the same time—each focused on a different aspect of how the virus propagates, such as the transcription of the viral code or its assembly—single mutations capable of circumventing one drug are nevertheless overwhelmed by the attacks of the other drugs, and no single mutation can ever gain enough of a foothold to allow the viral population to survive. In the language of landscapes, these drugs interact in a nonlinear way with one another. Each drug may fail on its own, but a cocktail of all of the drugs defeats the disease.

The HIV/AIDS cocktail is a triumph of modern science. Understanding the role of mutation and the various molecular mechanisms driving the disease required the costly efforts of thousands of researchers. Once this understanding arose, the strategy of simultaneously using multiple drugs, each of which defeated a different class of viral mutations, became an obvious approach.

While developing treatments based on a deep under-standing of the underlying molecular principles involved in a disease is desirable, such knowledge tends to be extremely costly to acquire. This cost restricts the number of drugs and drug cocktails that can be developed via this route. Occasionally new therapies arise by accident. For example, the drug cocktail used to treat Hodgkin's lymphoma was discovered by guesswork on the part of a rebel doctor who experimented on desperate patients. Still, as noted above, there are sound reasons to think that drug cocktails might be a good, and perhaps even necessary, way to treat dis-ease. Moreover, we already have a vast collection of chem-ical compounds—discovered by serendipity, hard work, or both—available for inclusion in drug cocktails. Thus we should be in a great position to develop new, cocktail-based approaches to curing disease. Of course, the difficulty with pursuing this scientific program is that drugs often interact with one another in surprising ways, and therefore we face a nonlinear landscape in which simple methods, like form-ing cocktails from those drugs that work well individually, are likely to fail. Fortunately, our knowledge of how to search rugged landscapes provides a potential way out of this conundrum.

In collaboration with Ralph Zinner, a medical doctor at MD Anderson Cancer Center, and some of his colleagues,

I have been using various search algorithms inspired by complex systems to find novel and effective chemotherapy cocktails. Our first test involved a set of nineteen therapeutic drugs that we were able to beg, borrow, and appropriate from fellow researchers. For each drug, we first found the dose that would typically kill off 10 percent of a particular strain of lung cancer cells, which we grew in plastic plates dotted by ninety-six wells, each roughly the size of a pencil eraser. There are more than half a million (2^{19}) unique cocktails that you can form from nineteen drugs each used at a fixed dose. Unfortunately, even with the heroic efforts of our laboratory technician and key collaborator, Brittany Barrett, we could test only about twenty cocktails every week, given the constraints of having to grow the various cell lines, mix the drugs, incubate the cells, and assay the outcomes.

To overcome these limitations, we introduced a search algorithm much like the hill climber described previously. We started with twenty randomly formulated cocktails. Each cocktail was added to a few wells growing the cancer cells, and after a few days we did an assay to see how many of the cancer cells survived. Cocktails that killed more cancer cells or used fewer drugs (with an implicit trade-off of including a drug in the mix if it killed at least 10 percent more of the cancer cells) were judged to have more fitness, a measure akin to our hiker's elevation. We took the fittest cocktail out of the initial twenty and made that our status quo. Then we started to hill climb. For each step, we looked at the nineteen drug cocktails that were identical to the status quo except for either including one drug that was not in the status quo or excluding one that was. Then we tested these nineteen variations against the status quo. If the fittest one of these variations was better than the status quo, we made that cocktail the new status quo and continued our search. If, on the other hand, when we looked at all of the variants we found that the status quo was still the fittest, the search was over, as we had found a local peak.

Within about nine steps, in which we investigated a couple of hundred cocktails out of the 524,288 possible ones, we found a three-drug cocktail that had a fitness more than four standard deviations above what we would have expected if we generated cocktails at random. Moreover, after discovering this three-drug cocktail, we did a literature search and found that two of the drugs had previously been suggested as being useful in a chemotherapy cocktail for another type of cancer. The third drug in our cocktail was not one we had expected to see, insofar as it tended to increase the growth of the cancer when used alone. However, in hindsight, there might be good reasons for including it. For example, the other two drugs might be particularly effective during cell division. If so, a drug promoting division might be a good thing to have in combination with the other two drugs.

This work was merely proof of a principle. It suggests that using the concepts of search on a rugged landscape may be a useful way to discover novel and effective chemotherapy cocktails.

In general, there are two serious challenges to discovering effective drug cocktails. First, nonlinear interactions among the drugs make simple search strategies, such as combining the best individual drugs, ineffective. Second, we face a combinatorial explosion of possible cocktails—that is, every time we add a drug (even at a fixed dose), we double the number of possible cocktails that we could test (namely, all of the previous cocktails with and without the new addition). With twenty drugs, for example, we have more than a million possible cocktails, with twenty-one drugs we have two million, with forty drugs we have a million million, and so on. Given this combinatorial explosion, we cannot feasibly generate and test all possible cocktails, even given modern advances in robotic laboratories. Most research gets around the combinatorial explosion by focusing on cocktails with only a couple of drugs—with twenty drugs, there are more than a million possible cocktails, but there are only 190 two-drug cocktails (if we ignore the ordering of the two drugs). Fortunately, search algorithms arising from the study of complex systems offer a possible solution to both of these challenges, as they are designed to find good solutions on nonlinear landscapes using only a limited number of experiments.

In many ways, the directed-discovery drug cocktail research above goes against various trends in medicine. A lot of recent work in chemotherapy has focused on drugs that target well-understood molecular mechanisms. While this is a useful approach, the effort it requires to uncover the underlying mechanism and design an appropriate drug is daunting. Directed discovery takes almost the opposite approach, even at times ordering the lab worker to make mistakes. An algorithm running in Santa Fe, New Mexico, with no knowledge of medicine or biology, takes a set of symbols and manipulates these based on feedback received from a laboratory in Houston (where a skilled technician mixes up the appropriate cocktails given the symbols she receives from the algorithm, adds them to a well of living cells, and after a few days of incubation reports back to the algorithm how many cells died in each well). While such a blind search may seem to be extreme, especially compared to the intensive, intelligent, and dedicated work of molecular researchers, ultimately what we really care about is finding useful cocktails, however they are discovered.

The directed-discovery approach to finding drug cocktails inhabits an interesting middle ground between intuitive leaps on the part of renegade doctors and the resource-intensive efforts needed to acquire deep molecular knowledge or to conduct large-scale screenings of natural compounds. Moreover, if directed discovery is successful, having information about what cocktails are effective may provide new insights into the underlying molecular mechanisms of the disease.

Notwithstanding the sound biological and complex-systems basis for pursuing drug cocktails, there are various institutional, legal, and regulatory constraints that favor single drugs. For example, drug companies like to focus their efforts on discovering (and patenting) a single, easily marketable, blockbuster drug, rather than seeking out cocktails that might involve drugs from other companies or component drugs that may have many substitutes. Even government regulations tend to favor single drugs. For example, the Food and Drug Administration currently requires that cocktails be tested and approved as a cocktail—an extremely costly process—even when the drugs that make up the cocktail are all individually approved. Recently there has been some promising movement on the part of such agencies to recognize the value of cocktails and encourage their use.

Over the coming decade we are likely to enter a new era of personalized medicine. For example, current cancer treatments tend to crudely classify types of cancers into overly broad categories and treat everyone within a category as if they shared the identical disease. Doctors run them through general protocols of treatment in the hope that something will work. The fact that people respond quite differently to the same treatment suggests that cancer is far more individually specific. For example, we know that there are various types of mutations in melanoma that make it more or less susceptible to different drugs. There is good reason to think that with further investigation and information, we will uncover many such specificities. As we enter an era of cheap genotyping, it is likely that future cancer diagnoses will be tied to the genotype of an individual's cancer. Once this occurs, we may enter a world in which each patient has what is essentially an "orphan" disease, shared with only a few others. In such a world we will need a way to quickly customize each patient's treatment. Directed discovery may prove to be a key enabling technology in such a future.

Ultimately, the notion of using cocktails to treat complex diseases is sound, and we need a systematic way to discover such cocktails given the innately rugged landscapes and combinatorial explosion underlying such searches. There are thousands of chemical compounds in the world that could be used to construct drug cocktails, and many of those compounds are quite inexpensive because their governing patents have expired. This storehouse of embodied chemical knowledge, linked with new developments in robotics and microfluidics, places us on the cusp of an era in which novel and effective drug cocktails, personalized to each patient, are waiting to be discovered—if, contrary to Six Sigma, we are willing to make some errors.

From Scarecrows to Slime Molds: *Molecular Intelligence*

> If you only had brains in your head
> you would be as good a man as any of
> them, and a better man than some of
> them. Brains are the only things worth
> having in this world, no matter whether
> one is a crow or a man.
>
> —L. FRANK BAUM, *The Wonderful*
> *Wizard of Oz*

BRAINS ARE OVERRATED, GENERALLY BY THOSE WHO HAVE THEM. Think (we didn't say brains weren't useful, they're just overrated) about all of the entities in the world that have to make good decisions with nary a neuron to be had. To take one example, each neutrophil granulocyte (a type of white blood cell, which you may have encountered in the form of pus emerging from a wound) in your body is capable of moving to sites of infection based on chemical signals. Once there, it can identify foreign microbes and destroy them by ingesting them or releasing antimicrobial chemicals. Such complex behavior abounds in every corner of our world. And it happens without neurons or a brain.

At one level, choice without neurons is not all that surprising. Even inanimate things, such as a drop of water or a rolling stone, have to decide where to go as they map out routes to low points in the landscape, and they do so in (at least ostensibly) clever ways. Indeed, countless computer hours are devoted to solving similar types of problems as we, say, try to find the shortest route for delivery trucks to follow. Of course, in the case of water and stones there is an external force, gravity, that drives the solutions—solutions that, despite our clever brains, we struggle to discover ourselves.

Water and stones, then, provide an existence proof that cleverness is not restricted to smart things. Yet the issue becomes far more interesting when we look at living things, which are making choices in arenas much closer to what we hold dear.

When Antonie van Leeuwenhoek perfected the microscope sufficiently to observe single-celled organisms, he noted that cells migrated in apparently purposeful ways. These observations were refined over the next few hundred years as various researchers realized that certain types of cells and organisms were directing their movements according to chemical signals in the environment.

The general name for this phenomenon is chemotaxis. To understand how it works, consider a lowly bacterium: *E. coli.* On its outside surface are several whiplike, semirigid, helical flagella. Each flagellum is attached to a chemical motor that can rotate it in either direction. When the flagella rotate counterclockwise, they all align into a single, corkscrew-like bundle, and in doing so propel the bacterium along a straight path. However, clockwise rotation causes the bundle to break apart, and the flagella flail about in all directions, causing the bacterium to tumble randomly. When one observes a bacterium, its motion alternates between these random tumbles and straight runs.

Although those may seem like limited behaviors, they are enough to get the bacterium where it needs to go. Suppose we place a drop of some chemical into the bacterium's world. The slow dispersal of the drop forms a chemical gradient with a high concentration where we placed the drop and a decreasing concentration as we move away from that point. Assume that the chemical we added is a nutritious food such as sugar. Once the drop begins to disperse, we will notice some interesting behavior on the part of the bacterium. While the bacterium still alternates between straight runs and tumbles, the time spent doing each action varies depending on the direction in which it is heading. When the bacterium is moving toward the drop, it tends to spend more time going straight than tumbling. When it is heading away from the drop, it tends to tumble more. This type of behavior will,

in general, cause the bacterium to move up the chemical gradient (see Figure 6.1). Similarly, if the chemical is
something that the bacterium wants to avoid, it will tend
to tumble more when it is headed toward the source and
less when it is headed away, and by doing so it will move
away from the area of highest concentration.

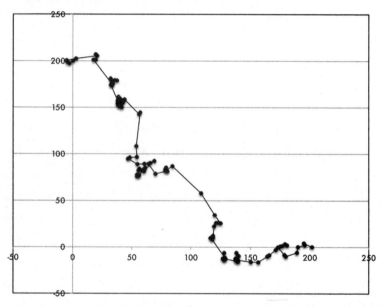

FIGURE 6.1: A simulated bacterium using chemotaxis to seek out a goal.
The bacterium begins in the upper left part of the diagram at coordinates
(0,200), and the goal is located in the lower right (200,0). The bacterium
alternates between tumbles (solid circles), which redirect it in a random
direction, and straight runs. The time spent in a straight run is proportionate
to the change in the number of molecules it encounters diffusing from the
goal—in the case of this simulation, it is controlled by a logistic probability
function of the distance to the goal from the old and new points. As seen in
the figure, this simple behavior is sufficient to direct the bacterium toward,
and have it remain around, the goal.

We know that the bacterium doesn't have any neurons or other obvious parts that could constitute a brain, yet somehow it is able to move toward good things and away from bad ones. In this case, molecules substitute for brains. While the exact molecular mechanisms and chemical reactions are a bit complicated (at least to an economist), in the end the bacterium is controlled by a complex system of molecules interacting with one another through the rules of chemistry.

When two molecules encounter each other, one or both of them may be altered. These alterations are tied to the physical shapes of the molecules, and there is a large amount of science devoted to understanding how chains of proteins fold into various molecular structures. At times, one molecule may "fit" into another, causing the receptor molecule to undergo a change. At other times, molecules can add or remove chemical groups to one another, roughly equivalent to turning on or off a chemical switch. Take these core abilities and mix in a few billion years' worth of evolution to tune various feedback loops and decay processes, and some rather sophisticated behaviors can emerge.

In the case of simple chemotaxis, the process is roughly as follows. When we place a drop of chemical into the environment, the enormous number (for argument's sake, let's say 6×10^{23} or so) of molecules it contains immediately begin to diffuse. Over time, this diffusion will form a gradient, with fewer and fewer molecules of the chemical as we move away from the initial drop.

As the bacterium moves around in its world, it encounters these molecules. The bacterium has receptors on the outside of its membrane that bind easily with the molecules—a common analogy here is to think of the molecules as keys that can attach to and open the lock-like receptors. When a molecule binds to the receptor, it causes a cascade of chemical changes inside the cell that lead to two important behaviors on the part of the bacterium.

The first behavior caused by the binding of, say, a repellent molecule is that the receptor releases a new set of molecules—to avoid confusion, we will just call these latter molecules signals—inside the cell. These signals lead to various cascades of reactions in the bacterium that propagate the initial signal and eventually cause the production of a new, short-lived signal to be sent that reverses the flagellar motor. Normally these motors run counterclockwise (at around 6,000 rpm), implying swimming in a straight line, but when the motor reverses, the flagella become all askew and the bacterium tumbles. The short-lived signal causes tumbling, after which the motor quickly resumes its normal rotation and the bacterium is back on the straight and narrow. If, instead of a repellent binding to the outside receptor, an attractant binds, then fewer such reversal signals are produced and the bacterium tumbles less. Thus, the molecular pathways induce a very adaptive behavior: when the bacterium is around attractants, it tends to go straight, but when it encounters repellents, it changes directions.

The second behavior caused by the binding of the outside molecule involves a feedback loop on the outside receptor itself. The sensitivity of the receptor is tied to the binding of external molecules *and* to some internal changes caused by such bindings. As more binding occurs, the receptor desensitizes itself to the outside molecules, essentially adapting—on a short-term basis, over the time span of a minute or so—to the outside level of the chemical. Thus, when an attractant molecule binds to the receptor, it both immediately reduces the number of tumbles (as previously discussed) and makes the receptor less sensitive to the attractant for a short while thereafter. If the bacterium meets a similar level of attractant over the next few minutes, it will tumble at its normal rate rather than the reduced one. This feedback mechanism allows the bacterium to "remember" its short-term past. Thus, if it finds itself in roughly the same situation it found itself in just a short while ago, it reverts to its normal behavior (occasional tumbles), and it alters this pattern only if it detects a change relative to the recent past. This memory induces exploratory behavior in the bacterium when it finds itself in the same situation as in the past.

Above, we focused on the case when the bacterium confronts either an attractant or a repellent, but what happens if it encounters both simultaneously? In such a case, the bacterium needs to decide what to do by making a trade-off between the potential benefits of the attractant and the costs of the repellent. This question was first explored by Wilhelm

Pfeffer in 1888, and he found that the relative strength of the two gradients is what matters—if the attractant gradient is stronger than that of the repellent, the bacterium moves forward, and if not, it moves away. Thus, the same molecular decision-making process that allows the bacterium to seek out good things and avoid bad ones is also capable of making trade-offs when both are present, giving the bacterium a set of preferences.

To move up a notch of behavioral complexity from our simple bacterium, consider the amoeba *Physarum polycephalum*, better known as a type of slime mold. During one part of its life cycle, this slime mold consists of a single-celled entity (with multiple nuclei, no less) that searches for food in an amoeba-like manner. Like our bacterium, this animal has no neurons, so its behavior must be dictated by various molecular pathways as well.

This slime mold enjoys food but dislikes light (which, among other harmful effects, disrupts its cellular processes). As a first step in understanding its decision making, two researchers in Australia, Tanya Latty and Madeleine Beekman ("Irrational Decision-Making in an Ameboid Organism," *Proceedings of the Royal Society B*, 2011), created an environment that contained patches that varied in the available amounts of food and light. Then, by introducing a slime mold into this environment, they could see which patch it preferred.

By mixing the conditions of the patches, they could derive the preferences of the slime mold—for example, does it prefer high food and light over low food and darkness? It turns out that a starved slime mold prefers a dark and high-food patch to a light and high-food patch, which is better than a dark and medium-food patch, and so on. A well-fed slime mold has slightly different preferences, preferring a dark and high-food patch to a dark and medium-food patch, which is better than a light and high-food patch, and so on. These and other tests reveal that the slime mold's molecular decision-making mechanism seems to be able to make reasonable trade-offs between various levels of good and bad things. Moreover, it demonstrates how those preferences can be influenced by the slime mold's internal state: when it is starved, it will trade off additional risk (in the form of being exposed to more light) for extra food.

This seems rational enough, but consider decision making in more complicated settings. It is well known that humans often violate what seem to be reasonable tenets of decision making. For example, suppose you give someone a choice between dining at a restaurant with good food and a crummy location (the Dive) or one with crummy food and a great location (the Tourist Trap). By tinkering with the food and location qualities between these two places, we can create a situation in which an individual is indifferent about whether she goes to either the Dive or the Tourist Trap. Now, add a third restaurant that is clearly inferior to the Dive—for example, the same type of location but slightly worse food. A priori, you might predict that such an addition will make no difference to the individual's final choice, since the new option is inferior to an already existing choice, and thus it should be immediately discarded as irrelevant. However, studies show that the addition of such an irrelevant alternative actually alters people's behavior, causing them now to favor the Dive over the Tourist Trap (see Figure 6.2). Marketers are well aware of this attraction effect and will often introduce an inferior product to boost sales of an existing one. By the way, there are other manipulations one can do, such as introducing another restaurant that is like the Dive but is a bit better on one dimension (say, slightly better food) and a bit worse on the other (slightly worse location), that will also alter decision making in unexpected ways (in this case, making consumers now prefer the Tourist Trap).

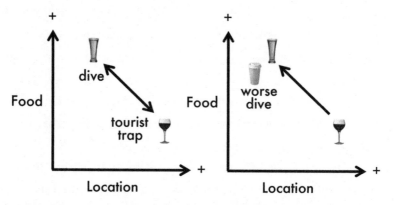

FIGURE 6.2: Two restaurants, the Dive and the Tourist Trap, differ across two dimensions in such a way that people are indifferent between the two (left panel). The addition of an irrelevant alternative (in this case, a restaurant that is inferior to the Dive on both dimensions) causes a shift in preferences to favor the Dive over the Tourist Trap (right panel).

It turns out that slime molds fall prey to the same decision errors that humans do. Start by creating two patches, one that is dark with little food and one that is light with more food, such that the slime molds choose each one about equally. Now introduce a third, inferior option (such as a dark patch with even less food than the preexisting dark patch). Even though this new patch is inferior to one of the existing patches and thus should be irrelevant to the two preexisting choices, we find that it causes the slime molds to gravitate to the preexisting dark patch over the one with more light and food (at least if they are not starved).

Not much is known about the mechanisms that underlie the slime mold's choices. It is likely that some molecular mechanism, perhaps similar in spirit to what we see in chemotaxis, is at play here. Regardless, both slime molds and bacteria, without a neuron to be had between them, can make productive decisions.

The ability to change behavior depending on the environment one encounters and to make (at least most of the time) appropriate trade-offs is crucial for survival. As complicated as this behavioral complexity seems, it's not hard to imagine how such mechanisms could evolve. Cellular mechanisms that take molecules from outside the cell, such as nutrients, and transport them into the cell are a fundamental part of life. Cells with more specific transport mechanisms that, say, move only particular types of molecules are far more likely to survive if the selective transport is beneficial. From such simple beginnings, it is easy to develop receptors that, instead of transporting specific molecules into the cell, just communicate the outside molecule's presence by emitting an intracellular signal. From here, it is not a great leap to have such signals induce productive internal actions (such as controlling the rotation of the flagella), cascade or degrade in useful ways, and even provide feedback to the initial receptor.

All things considered, the behavior we observe in these single cells is fairly sophisticated. Nevertheless, it is tempting to minimize the bacterium's accomplishment by claiming that because it is only the result of a set of straightforward molecular reactions, it is therefore trivial and should not qualify as any real form of thinking. Surely brains composed of neurons must provide a deeper form of thinking.

But what is so special about neurons anyway, other than that humans tend to have more of them than most other species? What neurons are really good at is transmitting signals across large distances. Our bacterium is relatively tiny, so molecules floating around and randomly bumping into one another are sufficient to transmit signals, as even randomly drifting things bump into one another quite often on such small scales. However, as an organism gets bigger, it needs a more reliable and rapid mechanism to convey signals across large distances—"when it absolutely, positively has to be there," you need something like a neuron to quickly transmit and connect signals. Other than the necessities of scale, there may not be much difference between these systems. Indeed, the fact that molecular decision mechanisms lead both to rational decisions and to errors similar to the ones humans make with fully developed nervous systems suggests that the two systems may be driven by similar principles.

If that is true, it may be the case that thinking exists on a much broader scale than we normally imagine. If simple molecular mechanisms allow bacteria, slime molds, and their lot to incorporate fairly sophisticated behavior, we might want to examine other systems of signals and reactions to see if they, too, embrace thinking. Perhaps any set of interacting molecules can be interpreted as performing a thinking-like computation.

Moreover, if signals and reactions are all that are needed for thinking, then brains may not be restricted to individuals. Perhaps larger-scale social systems, such as honeybees in a hive, people in an organization, or a collection of interconnected markets, may be performing thinking-like computations. We'll explore this topic in the next chapter.

We likely are surrounded by brains everywhere, some of which we easily recognize and admire and others of which we are only beginning to understand.

SEVEN

From Bees to Brains:
Group Intelligence

> For so work the honey-bees, creatures
> that by a rule in nature teach the act of
> order to a peopled kingdom.
>
> —WILLIAM SHAKESPEARE

CONSIDER FOR A MOMENT HAPPENING UPON AN ALIEN ORGANISM. It has an outer membrane that contains its insides while also keeping the outside world at bay. When potential predators approach, it has the ability to send out poisonous projectiles to repel them. Inside the organism, tens of thousands of particles perform a variety of functions, including maintaining this apparently warm-blooded organism's interior temperature within narrow bounds, transporting waste products to the outside world, creating new particles and maintaining old ones, and many other internal processes necessary for survival. The organism can send out into the world long, thin tendrils that grope randomly about, apparently in search of nutrients. Once a tendril happens upon a rich source of nutrients, the organism sends out additional tendrils that go directly to this location to help transport back the find. Incoming nutrients are converted by various processes into a form of storable energy that can be used to keep the organism alive (much the way our own cells convert molecules such as glucose into ATP). Finally, we find that there is one specific particle in the organism that appears to serve a role akin to our own DNA, as it contains the needed information to create a constant stream of the other particles critical to the organism's vitality.

As we continue to observe the organism over the course of a year we notice an interesting phenomenon. Typically during the spring, the organism appears to erupt about half of its particles, which then settle in a nearby location. Tendrils emerge from this mass, too, but rather than seeking out nutrients they seem to be searching for a new outer membrane—much the way a hermit crab seeks a new shell. As we watch, the tendrils connecting the mass to potential membranes either strengthen or wither away, until eventually only one strong tendril remains. At this point, it is as if this tendril pulls the entire mass of particles into its new skin. Through this process, what started out as a single organism has divided into two. These two newly constituted organisms return to the necessities of survival, seeking out new sources of nutrients, growing in strength, and, if conditions remain favorable, dividing anew.

The organism described above is not some strange new form of extraterrestrial life but rather a colony of honeybees viewed from enough distance that we have difficulty making out the individual bees. This new point of view suggests that it may be useful to think of a hive of honeybees as a single superorganism, rather than as tens of thousands of individual bees.

How does such a superorganism function? When we normally think of a hive, it is all too easy to view it as a monarchy ruled by a benevolent queen, who directs her workers on the daily tasks of gathering nectar and pollen, processing and storing honey, and all of the other duties that keep the hive alive. Alas, this comforting story of a finely tuned aristocracy could not be further from the truth, though the reality here is far more rich, fascinating, and useful.

As we saw in Chapter 6, even collections of molecules interacting via a fixed set of chemical rules can display intelligent behavior, like a bacterium seeking out good things and avoiding bad ones. Given that molecular interactions can provide a cell with the ability to make intelligent decisions, then perhaps interactions of other types of particles, here honeybees, could bestow intelligence on those systems as well. As we will see, what's true of honeybees will help us unlock the mysteries of other systems of decentralized, interacting particles that are making intelligent decisions, from molecules in cells to neurons in brains, bees in colonies, traders in markets, workers in firms, and beyond.

HONEYBEE COLONIES, WHILE VIBRANT DURING THE SPRING and summer, are often on the edge of survival during the winter as temperatures drop and nectar ceases to flow. To successfully overwinter, the hive must have enough workers to maintain its temperature and to rapidly repopulate the colony when the nectar begins to flow in the spring, but too many workers will exhaust the limited stores of honey before winter ends. Thus, colonies must carefully control their size.

Swarming typically occurs in the spring, when nectar is plentiful, honey has been stored, and the colony is rapidly growing. When the colony swarms, the old queen takes off with roughly half of the hive. The exiting workers take their share of honey, and the resulting swarm alights nearby (see Figure 7.1). The swarm, with the queen huddled in the middle, is vulnerable at this time, as it is exposed to the vagaries of weather and predation. As the swarm settles in, a few scouts head off and search for potential nesting sites—perhaps a hollow in a tree or a cavity in a building. When a scout finds a possible site, she explores it and makes an evaluation of its overall quality. (Through a clever set of tedious experiments, we now have a good sense of the qualities that scouts are seeking in a new home. In particular, they want some combination of a properly sized cavity,

FIGURE 7.1: A swarm of wild honeybees that formed under the eave of a building just down the hall from my Carnegie Mellon office. After many days of an unsuccessful search for a new home, the swarm began to build the comb in this exposed location. After a few weeks the swarm disappeared. *(Photograph by the author.)*

a certain height above the ground, and an entrance of a certain size and orientation.) Once a scout explores a particular site, she returns to the swarm and begins to communicate the location to the other scouts by way of a waggle dance performed on the outside surface of the swarm. The key to this step of the process is that the time spent dancing is tied to the scout's perception of the quality of the site, with higher-quality sites receiving longer dances. Other scouts on the swarm observe the dancing and are induced to check out the advertised location for themselves.

It is at this juncture that we get our first hint about how such a decentralized system could possibly work. Since the sites that are perceived to be better receive longer dances, and since new scouts are recruited by observing the dances, it is more likely that recruits will be sent to the potentially better locations. This results in a positive feedback loop for the higher-quality sites. Even with this positive feedback, there is a subtle mechanism built into this system that prevents locking into a bad choice too quickly. Since potentially better sites are explored more often, they receive many more evaluations of quality, so even if the initial scouts are somehow poorly calibrated about the site's true quality, subsequent investigations will tend to correct such errors.

Imagine observing the outside of the swarm and tracking what locations are being advertised by the dancers. When the swarm first alights, we see no activity other than scouts heading off in random directions. As the first scout returns, she dances for whatever location was uncovered, and perhaps this induces some other scouts to go see it for themselves. Soon other scouts return from their initial forays, and new locations are advertised. Over time we start to see several sites being advertised, and we could even track a site's popularity by counting the number of dancers over some given time interval. Like the *Billboard* music charts, we see some sites that have been charting for a few periods, with new sites occasionally breaking into the mix rising with a bullet and, at times, once popular sites falling from grace like one-hit wonders, never to be heard from again.

This mechanism allows the swarm to conduct many parallel searches and to find a new home relatively quickly. There is no central accountant noting each scout's findings and deciding where to send the next scout or when to end the search and move. Rather, scouts are directed by their own, local observations of what is happening in their small neighborhood on the swarm's surface. The system has various indirect checks and balances, such as more intensively investigating the more promising sites while also maintaining a variety of other options.

After a day or two of dances, one location begins to emerge as the favorite (see Figure 7.2). At this point, the decentralized system has essentially made its choice.

How is this ultimate choice finalized and communicated to the swarm? While it might be possible for all of the bees to somehow sense that the dances have converged, this strains any reasonable notion of a decentralized system governed by only local communication. It is here that the second key element emerges in our story, one that has the simplicity and pure functionality of a Shaker chair. Rather than being driven by any action on the surface of the swarm, the final decision is made at the discovered site. Studies have shown that scouts exploring a site have the ability to sense when a quorum of around twenty bees has formed, and it is the occurrence of such a quorum that triggers the final decision. Once a quorum is present at the new site (note that sensing such a quorum is not all that difficult given the information locally available to the congregating

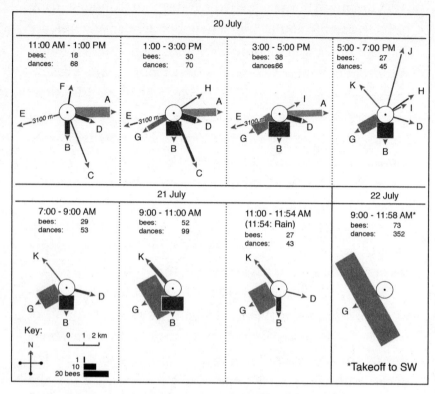

FIGURE 7.2: A time series of panels representing the number of honeybees dancing for various hive locations on the surface of an actual swarm. Each arrow represents the direction and distance to a potential hive site, with the width of the arrow indicating the number of dances for that particular site. The swarm identifies eleven different sites during the search. Eventually the swarm begins to concentrate on sites B and G, and after a pause due to rain on the second day, site G becomes the consensus choice. (The data and figure are courtesy of Thomas Seeley.)

scouts), all of the scouts return to the swarm and begin to make specific noises (known as piping) and perform "buzz runs" that are akin to some crazed motivational speaker running through an audience. This causes the honeybees in the swarm to warm up their flight muscles and get ready to make the big move.

All that remains is the issue of how a few tens of bees that know where to go can lead a few thousand that do not. The answer to this quandary turns out to be that a handful of well-directed and fast-moving bees is sufficient to direct the large, slower-moving swarm to its new home.

Russell Golman, David Hagmann, and I have modeled the decentralized decision system above in a simple way. Consider an urn filled with one differently colored ball for each potential site. We mix these balls about and then blindly reach in and pick one. Whatever color we pick is the option we investigate, and we then place the chosen ball back in the urn and also add more balls of the same color. The number of additional balls depends on the "quality" of the color that we picked. For example, suppose we have four different colors in the urn and whenever we pick, say, a red one, we put it back with two new red balls (similar to how a scout does a longer dance after visiting a good site), while whenever we pick any of the other colors, we replace it and add only one new ball of the same color. After we replace the balls, we mix the balls up again and draw anew.

This urn process is similar to what happens in the honeybee system. At any stage of the process, the likelihood of picking a particular color (potential site) depends on the proportion of that color in the urn. When we first reach into the bin, we have an equal chance of picking any color. Once we make our first choice, the scheme for replacing subsequent balls has an effect similar to the dancing scouts, as the better option (here, the red ball) gets more balls added back into the urn than the other colors. This increases the likelihood that the better option will be chosen in the future. The system makes a choice as soon as the number of balls of a particular color equals or exceeds a preset quorum level.

The behavior of this model is shown in Figure 7.3. Since we start with one ball of each color, if the quorum is two, then whatever color ball that we pick first will become the final choice, as the addition of a new ball (or two) of that same color will meet the threshold. Thus, with a quorum of two, there is an equal chance (25 percent) of choosing any of the four colors. As the quorum increases, the differential addition of balls based on color (the best getting two, the others getting one) begins to matter more and the system becomes more likely to pick the best choice. For example, at a quorum of five, the system picks the best choice about 50 percent of the time. At a quorum of twenty, there is a 70 percent chance of picking the best option. We can prove mathematically that as we increase the required quorum, the system is more likely to pick the best option (and as we allow the threshold to increase without bounds, the system always converges to the best option).

There is an important trade-off in this decentralized system: as the required quorum level increases, so does the time

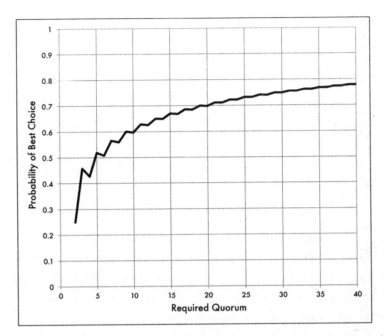

FIGURE 7.3: Likelihood of a quorum forming for the best choice (*y*-axis) with four different colors in the urn given different quorum thresholds (*x*-axis). Here, anytime the best color is chosen we add two additional same-colored balls to the urn, while choosing any of the other colors results in the addition of only one same-colored ball. As the required quorum size increases, the likelihood of picking the best choice increases, but so does the time that it takes (not shown) to reach that quorum.

needed to hit that quorum. Thus, if we want the system to make better choices, we need to wait longer. Typically, waiting is costly, especially in a system such as our honeybee swarm, which is quite vulnerable to the elements and predators and which cannot, in its current configuration, make and store honey. Swarms usually have only a few days to find a new home before they are so compromised that their continued survival is unlikely. Thus, waiting too long may be worse than picking an inferior choice. Given this, we might expect that evolution would result in required quorum levels that satisfice—that is, that cause bees to choose a reasonable, even if not optimal, home rather than wait too long. It appears that swarms use quorum thresholds of around twenty, and given our model, this results in a mechanism that works relatively quickly with a good, but not perfect, chance of finding the best home.

Another interesting feature of the above decentralized decision-making system is how it weighs risky choices. A typical way to derive the risk preferences of decision makers is to give them choices between a safe alternative and an equivalent (in terms of expected value) risky one. For example, suppose you have a bet that on average pays $2. You have a choice of how you can get that average value, however: win $2 for sure or take a gamble that 50 percent of the time pays $1 and 50% of the time pays $3. Which would you pick? If the second gamble had an 80 percent chance of paying $1 and a 20 percent chance of paying $6, would your choice change?

In Figure 7.4 we show how our urn makes such choices. The system must pick between receiving two additional balls for sure or one of the risky gambles shown in the legend on the right. While all of the gambles have an expected value of 2, the variance—a proxy for riskiness—of the gambles decreases as you move from the top down. The graph shows the likelihood of picking the safe choice on the y-axis, so

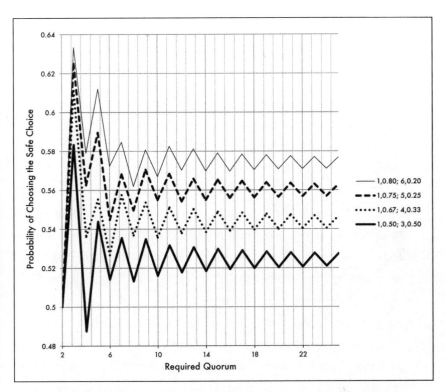

FIGURE 7.4: Probability of choosing the safe choice (2 for sure) versus the risky choice (various probabilities of receiving either 1 or a higher value than 2, such that the expected value of the gamble is always 2—for example, in the topmost gamble, there is an 80 percent chance of receiving 1 and a 20 percent chance of receiving 6). As long as the required quorum is greater than 2, then (with one exception) the system is more likely to take the safe choice instead of the risky one, since the probability for that outcome is over 0.50. As the gambles become more risky (the legend on the right is ordered from most risky at top to least risky at bottom), the system displays increased risk aversion by choosing the safe option more often.

anytime this is above 0.50 the system is risk averse. As soon as the threshold exceeds 2, the urn system is more likely to choose the safe choice over the risky one in all but one quirky case, and as the variance of the gamble increases, the system prefers the safe choice even more. Thus, the decentralized decision system used by the swarm, insofar as we can capture its behavior using our urn, is risk averse.

Risk aversion may be an important strategy in evolutionary systems, since a single failure of a species to reproduce will prune its branch off the evolutionary tree. You are the result of a *very* long chain of successful matings going all the way back to the origin of life. One broken link in this chain and you would not be here. Of course, there is a difference between you not being here and *Homo sapiens* going extinct, but the general idea still holds. An evolutionary strategy that relies on gambles versus sure things is, literally, quite risky. Under some circumstances, playing the odds could potentially make sense, but the huge cost of failure—extinctions, like diamonds, are forever—may restrict such a strategy to very particular conditions. Often, slow and steady wins the evolutionary race. In the case of a honeybee hive, there is a tremendous investment involved in cleaving the hive. Given this, evolutionary success comes from finding a reasonable home for sure rather than taking a gamble that gives the bees a small chance of a much better home and a large chance of a much worse one.

Honeybees are not unique in having a decentralized decision-making process. Some species of ants have a related approach to finding new homes. When a colony of these ants decides to move, it sends out scouts to search for new possibilities. The speed at which scouts that have investigated a possible site return to the original colony to recruit new ants is tied to the quality of the new site—better-quality sites cause the scouts to return faster.

When a scout returns to the original colony, it teaches another ant the way back to the new home by a process called tandem running. Tandem running requires the tedious training of the recruit, but it has the advantage that the recruit learns the needed landmarks along the way so that it too can return to the original colony and teach others. Like the bees, when a quorum forms at the new site, there is a change in behavior. Rather than using tandem running, the ants returning to the original home just pick up their nestmates and carry them to the new home. The advantage of carrying is that it is about three times faster than tandem running. The disadvantage is that it doesn't allow the new recruits to learn the path, but such knowledge is not needed once the new nest is chosen.

Do we see hivelike minds forming in human systems? Perhaps. Think back to the analogy to the *Billboard* music charts. As songs get discovered, they get played more often. As they get played more often, more people hear them, and if they like them, they start to play the song as well (or, more likely, listen to stations that play the song more often). Over time, the most popular songs rise to the top and can often remain there for long periods of time. Low-quality songs fail to recruit new listeners and fall off the charts, never to be heard again.

Consumer goods, such as MP3 players and smartphones, may follow similar processes. At the start of these markets, early-adopting consumers go out and buy on a whim. Every time these adopters use the product, they are essentially performing a waggle dance for the other members of the hive. The better the product, the more likely they are to use it—that is, the longer the dance. As new consumers enter the market, they observe others, and those observations influence their own purchasing behavior. Note that having a product with distinctive observable features (think of the iPod's white earbuds) is an advantage here (and especially in this case, since the main bulk of the product is usually hidden away in a pocket). As the positive feedback loops begin to mount, we often see one product take off and begin to dominate the market.

Another example of human social systems undergoing hivelike processes arises in political contests. In primary races, we have numerous candidates vying for recognition. Candidates get their buzz from contributions and the loyalty that they engender from potential voters who show up at rallies and promote the candidate to their friends and neighbors. Of course, there are forums such as debates, where candidates get exposed to large swaths of the citizenry, but even during these events support from the audience (via, for example, applause) or from the moderator (by asking more questions of the front-runners) influences the views of others. In the early stages of these races, front-runners come and go as political fortunes rise and fall. Over time, though, a clear front-runner emerges and the nomination becomes solidified. (I'll leave the question of whether this person represents the best of the bunch for another time.)

While these examples focus on large-scale social choices, we can take the same ideas and look inward to the choices that each of us makes.

When we considered the honeybee hive, we ignored the fact that the behavior of each scout bee was directed by a brain with around 1 million neurons (by comparison, ants have from 100,000 to 250,000 neurons, and we have around 11 billion). Instead, we abstracted away the issue of each bee's behavior being controlled by a bunch of neurons and just considered each bee as being a particle following some simple rules. By doing so, we were able to gain insight into the behavior of the swarm as a whole.

One of the hallmarks of the study of complex systems is that the particulars of the system's behavior at one level, say, the neurons in a honeybee's brain, can be treated abstractly at another level, say, by introducing the assumption that individual bees follow simple rules, regardless of how such rules are generated. By making such abstractions, we can then focus our investigation on the current level of behavior, which in the above case is how interactions of individual honeybees allow the swarm to make a unified choice about where to relocate. When we take this journey of abstraction and move either up or down a level, we can (we hope) apply the same general principles we've uncovered at one level to another. Perhaps the ways that interacting atoms create molecular behavior, interacting molecules create chemical behavior, interacting chemicals create neuron behavior, interacting neurons create individual behavior, interacting individuals create colony behavior, and interacting colonies create ecosystem behavior are all governed by similar principles.

If we are lucky, insights from one level will seamlessly flow into our understanding of what happens at another level. This offers the possibility that there is a deep similarity among such systems. If so, maybe the insights we gained from studying how the decentralized interactions of honeybees result in swarm choice can be used to understand the choice behavior of other systems. Perhaps we have come upon a simple way of understanding choice making ranging from bees to brains.

One advantage of observing a swarm of honeybees is that we can see all of the honeybees and track their individual movements. Of course, even in this case doing so is not trivial, as you have to label and track thousands of honeybees, but it's still easier than tracking the billions of neurons that can make up a brain. While we can't track billions of neurons, we can insert extremely thin probes into a brain and observe the activity of a single neuron. From many such observations, we begin to understand how the behavior of individual neurons results in a collective decision.

William Newsome and his colleagues have been using this technique to analyze how macaque monkeys make decisions. For example, they present a monkey with a screen of randomly placed and moving dots, with some set proportion of these dots all moving either to the left or to the right. The monkey is trained to decide whether the majority of the coherent dots are moving left or right and to indicate this decision by moving its eyes to a particular spot on the screen.

Such a decision is possible because within the visual cortex of a mammal's brain are highly specialized neurons that can detect some remarkably specific features of what is being viewed. For example, certain neurons fire only when the eye sees, say, a horizontal edge. Other neurons specialize in detecting motion in a specific direction. It is the firing of these motion-detecting neurons that is the crucial sensory input into the decision that Newsome's monkeys were making. Recall that only a proportion of the dots are moving coherently, so the input signals from these neurons are rather noisy. In another area of the brain, a different set of neurons begins to weigh the incoming sensory evidence. These neurons track how much the left-favoring motion neurons are firing over time compared to the right-favoring motion neurons, and when the total amount of observed firing in one of the two directions begins to dominate, a decision is made. When a lot of the dots on the screen are moving in the same direction, the decision is easy, quick, and free of error. When the signals are more mixed, such as when there is only a small proportion of coherent dots, the decision takes more time and is less accurate.

The accumulation of evidence in support of one position or another, eventually leading to a final decision, is similar to what we see in honeybee swarms. In both systems, evidence is slowly accumulated over time to the point where it becomes so overwhelming that a final decision is made. In the honeybee system, as we have seen, safeguards are built into the process that result in more testing of the better-appearing choices. It is not immediately obvious that the brain has an analog to this, though it is possible that firing neurons might either atrophy or alter the sensitivity of related neurons, resulting in similar behavior. Even without this, the similarities between what might at first appear to be rather disparate systems is compelling—bees and brains may be closely related.

There is no guarantee in any of these systems that the final decision will be the right one. It is possible for the honeybee swarm to choose an inferior hive, perhaps because the best option was never identified in the first place, or because, given chance events, it was not sufficiently reinforced while an inferior choice was able to get enough of a foothold that the positive feedback worked in its favor. Similarly, random events in the brain can cause enough noise in the firing patterns of the motion neurons or errors in the decision neurons to reverse the choice. The monkeys get about 95 percent of the decisions correct when 51 percent of the dots moved coherently and only 70 percent correct when 13 percent were coherent.

The similarity of mechanisms between bees and brains suggests that perhaps a deeper connection across systems can be made. It may be that a lot of systems that seem intelligent are in fact the result of simple, interacting particles. We know that complex systems are adept at taking groups of interacting particles and forming larger-scale structures that are no part of any particle's intention or individual ability. Thus, perhaps having larger-scale structures display some extraordinary intelligence is not too surprising.

An ant colony, like a bee colony, has to make a variety of decisions, such as whether to send out workers to find food versus repairing the mound and so on. Deborah Gordon and others have found that when an ant decides what to do, it is influenced by the actions of other ants. If an ant encounters a lot of other ants returning with food, she too will go out and gather food. If food is plentiful, it will be easy to find and ants will return faster with food. That will encourage other ants to seek food as well. If food is scarce or if there is a predator about, few ants will return with food, and workers will do other tasks. In either case, the rule that says "Do what other ants are doing" leads to productive behavior for the colony.

That is not to say that blindly following a rule will always be optimal. For example, army ants follow the chemical signals laid by others. This behavior is usually adaptive, as it organizes the actions of tens of thousands of ants into formations suitable for quickly moving the colony or hunting prey, based only on local signals and without the need for global directions. Unfortunately, such a strategy can sometimes fail when a line of army ants inadvertently begins to follow its own trail, forming a circular mill (see Figure 7.5) that, with time, ends badly for all involved.

Global behavior is the necessary result of any set of interacting particles. Sometimes this behavior is chaotic, disorganized, and hard to fathom, as in the molecules of air that surround you, bumping off one another like pool

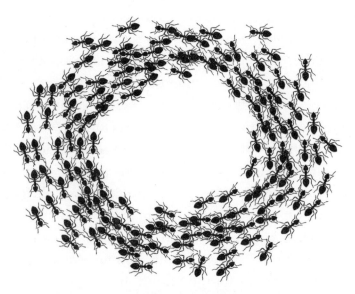

FIGURE 7.5: A circular mill of ants forms when the simple rule governing each ant's behavior—namely, follow the pheromone trail left by other ants—goes awry, and they accidentally begin to track themselves. Within a couple of days, ants caught in the mill will perish.

balls—although even in this kind of system there is an average behavior that, with good reliability, distributes molecules roughly evenly about the room instead of having them all concentrated in one corner. (This latter behavior is possible, though not very likely.) At other times, local interactions result in global behavior that appears far more organized and coherent. There is no guarantee that such organized behavior is productive and useful, though we do have examples where this is so. Army ants can form sixty-foot-wide, three-foot-thick fronts that go through the forest seeking prey like a bulldozer blade. Honeybee swarms make life-and-death decisions about where to relocate tens of thousands of bees. Neurons sense the outside world and formulate useful decisions about what action to take.

Such productive, self-organizing systems can be honed by evolutionary forces. While the laws of physics may be fixed, the internal chemical environments in which these play out are not. Thus, by recombining molecular soups of one type or another, different rules can be invoked, and if a particular set of rules results in global actions that lead to some higher purpose, evolution can capture these conditions for future generations. As a result, evolutionary forces can form organisms relying on complex molecular interactions that allow metabolisms, information processing, reproduction, and even thoughtful decision making by individuals and superorganisms.

MUCH AS SUCH SYSTEMS CAN EVOLVE, THEY CAN ALSO BE created and honed by humans. An interesting case of human-derived, local rules being designed for productive global outcomes arises in auction markets. Auctions are composed of simple sets of rules that every participant must obey. The obvious goal here is to find sets of rules that, even with self-interested individuals, will result in a sequence of trades that has some nice global properties, such as ensuring that the most profitable deals possible are made by the participants during the auction.

The earliest auction on record is from around 500 B.C. in Babylon. Women were auctioned off for marriage, with the revenue acquired from the sale of the most desirable wives being used to subsidize the trades of the less desirable wives. Such a mechanism reemerged in the 1990s in discussions of "feebate" systems. In these auctions, fees imposed on less-efficient technologies, such as a gas-guzzling truck, are used to fund rebates that subsidize the purchase of more energy-efficient vehicles.

Since Babylonian times, hundreds of different auction mechanisms have been developed, though only a few types are widely used. When most people think of an auction, it is the open-outcry English auction that comes to mind. Participants in this auction increase the bid until no one wants to bid any higher, at which point the good is sold to the highest bidder at the last price bid. In a Dutch auction—used to sell, among other goods, large quantities of freshly cut flowers each morning in Holland—prices start out well above what anyone would want to pay for the good. The price is then lowered until a buyer agrees to take the offered lot. Some auctions, like those commonly used in financial markets, combine features of both English and Dutch auctions, with potential buyers making ever-increasing offers to buy (bids) while potential sellers make ever-decreasing offers to sell (asks) until someone agrees to accept one of the currently available offers. Other auction mechanisms alter the rules in interesting ways. For example, in a second-price or Vickery auction, the potential buyers covertly submit their bids, and the highest bidder wins the auction—but the price she pays is determined by the second-highest bid instead of her own. A variant of the Vickery auction is used to sell billions of dollars' worth of United States Treasury bills each week.

The rules of an auction—created by the ingenuity and greed of man and the trials of history—are much like the behavioral genes that arise in the honeybee colony. Individuals, interacting through these rules, create global behavior that may be disconnected from the rules that have been imposed or even the goals of any individual. Some sets of rules lead to bad outcomes where, say, buyers or sellers are able to collude and take advantage of one another, or goods do not flow in a timely manner to their best use. Auction institutions associated with such bad outcomes tend to die out. Other rules, such as the English auction, persist over time, as both buyers and sellers find value in participating in such auctions. Moreover, societies that embrace these rules tend to thrive.

Auctions are just one example of how a human social system can adopt a set of simple rules designed to generate spontaneous and productive social order. Of course, many other examples exist. *Robert's Rules of Order* (first published in 1876) was modeled on the rules being used in parliaments and legislative houses, with the goal of directing the interactions of individuals in groups in a way that would result in the emergence of productive, group-level decisions. Similarly, constitutions, rules of law, courts, and international treaties are all derived in the hope of emergent wisdom, though even the best-laid rules can lead to a circular mill at times.

The theory of complex systems is still in its infancy. We know that there may be many sets of rules that lead to similar emergent properties, and thus one might suspect that this is also true for the types of systems discussed above. Take, for example, the notion of a democratic system. Since the time of the Greek city-states—at roughly the same time as the Babylonian auctions—a variety of democratic rules have been tried. They often differ from one another in the amount of representation and freedom given each citizen, yet a similar sense of democracy emerges from all of them.

Alternatively, consider the various tenets contained in religious systems, all trying to invoke a set of beliefs that will result in a productive society. Different religions try to invoke such beliefs in different ways, and even within a

given religious branch there are often refinements (come-
dian George Carlin was able to reduce the Ten Command-
ments down to just two: "Thou shall always be honest and
faithful" and "Thou shall try really hard not to kill anyone"), yet they all may lead to similar outcomes.

Having a better theory of emergent organization would
have a lot of practical benefits, allowing us to generate or
simplify sets of rules that will result in productive ends.
Think about the value of such an approach for something
like our tax code. In the United States, the current tax code
contains around 3.4 million words (the equivalent of about
twenty-four megabytes of data). These words create a set
of tax rules that, for better or worse, organize key parts of
our society, including government spending, income in-
equality, employment opportunities, industrial production,
investment options, political affiliations, the likelihood of
cheating on taxes, and on and on. The complexity of the
current system is very high, and perhaps unnecessarily so.
A theory of emergent organization might point to a vastly
simplified set of rules that could produce better outcomes.

Even without a full-blown theory of emergence, examples such as honeybees searching for a new hive location
may provide some useful insights. Evolution has enabled
the honeybees to discover a good home without relying on
centralized information or authority. Similar problems exist
in social, government, military, and business domains, and
perhaps these problems could be solved using related mechanisms. Various engineering problems may also be amenable

to such solutions, and honeybee-based mechanisms could be used to, say, create decentralized decision systems that gather and highlight key information ranging from web-based searches to intelligence gathering for business or national security.

While trying to modify Shakespeare for scientific accuracy is a fool's errand, nonetheless we'll try: "For so work the honey-bees [and brains and societies], creatures that by a [simple] rule in nature teach the act of [self-organized] order to a peopled kingdom [of scientists and practitioners]." While the poetry of the language is seriously compromised by these modifications, the poetry of the science is not. Governed by simple rules, interacting systems can result in spontaneous, system-wide behavior that is both a part of those underling rules yet wholly disconnected from them. Such magic can, and does, happen at all levels of our world, from bees to brains and beyond.

EIGHT

From Lawn Care to Racial Segregation: *Networks*

Conspicuous consumption of valuable
goods is a means of reputability to the
gentleman of leisure.

> —THORSTEIN VEBLEN, *Theory of the Leisure Class*

A T THE HEART OF ANY COMPLEX SYSTEM IS A SET OF INTERACTING agents. If we track who interacts with whom, we can uncover a network of connections among the agents. Not too surprisingly, the structure of these networks matters, both in terms of what types of networks exist across various complex systems and in terms of how different network structures influence system-wide behavior.

Consider a lake surrounded by houses. Each house in Lakeland is on the water, so for any given house there is only one neighbor to its left and one to its right. From a bird's-eye view, each house occupies a bit of space on a circle formed by the lake's edge (see Figure 8.1).

As in most neighborhoods, the behavior of each resident is influenced by her neighbors. To take just one example, suppose that each resident has to decide how much effort to spend on her lawn—say, whether to mow or not. The amount of effort that one exerts here may depend on the actions of one's neighbors. If the neighbors keep immaculate, putting-green-like lawns, then you might be

FIGURE 8.1: The community of Lakeland. Houses are arrayed around a lake, with each household interacting with its immediate left- and right-side neighbors.

inclined to do so as well. If the neighboring lawns resemble weed-infested jungles, then your lawn care efforts might wane.

To explore this world, let's assume that every Sunday each resident decides whether to mow her lawn. This decision is strongly influenced by her two immediate neighbors (one to the left and one to the right). To keep things simple, we assume that if both neighbors took the action opposite of what she did last week, then she will alter her action this week. Otherwise she will continue to do what she did the prior week.

This rule of behavior is equivalent to a crude form of majority rule. There is a group of three (the resident and her two neighbors) that is "voting" on what to do. If the resident and at least one of her neighbors did the same action last week, then this majority decision dictates what the resident does this week. If, instead, the resident deviated from both of her neighbors last week, then their two votes overrule hers, and she alters her behavior.

We have almost enough elements in place to begin exploring the system-wide behavior inherent in Lakeland. The one remaining piece is what happens during the first week of the lawn care season. The behavior above is predicated on the previous week's behavior, and obviously there is no previous week at the start of the season. So to initialize the system, we will flip a coin for each resident to determine her initial action.

At first glance you might think that, given majority rule, whatever choice is in the majority the first week will dictate the behavior for the second week, and everyone in Lakeland will either always mow their lawn or never mow it. While this seems intuitive, recall that the behavior of each resident is tied only to that of her immediate neighbors, so there is no way for the global information about the initial majority choice across *everyone* in Lakeland to be instantly transmitted to each resident during the second week. Given this observation, you might modify your initial intuitions and imagine that over time, as neighbor influences neighbor, the initial majority will slowly flow around the lake in such a way that the system eventually ends up, after a few extra weeks, coordinating on whatever majority decision was initially drawn. Alas, as in most complex systems, such sensible intuitions are wrong.

Suppose, for whatever reason, two next-door neighbors start to take the same action. If this occurs, each of these two residents will always have at least one immediate neighbor taking the same action that she is doing. Given majority rule, this implies that neither of these two neighbors will ever change her action in the future.

Thus, anytime two neighbors take the same action, they will lock themselves into that action for the rest of the season. Since this lock-in depends on the action that is common across the pair, it suggests that as we watch the system over time we will see the formation of islands of neighbors taking a common action (either always or never mowing).

For the moment, focus on the edge of one of these islands. If the nearest neighbor next to the island's edge ever decides to take the same action as the island, then that neighbor becomes part of the island, since she will always have at least one neighbor (the one next to her on the previous edge of the island) taking the same action as she is, and hence she will never want to change her action for the rest of the season. Over time, we might see the various islands slowly accreting new members as they absorb like-actioned nearest neighbors.

Thus, part of the dynamics of this system is a set of isolated islands of common action being established as pairs of neighbors happen upon the same action. At the start of the season, these islands will be scattered about the circle, with the exact location of, and common action for, each island being tied to the random initial conditions. Once established, these islands are likely to grow in size as they accrete like-actioned neighbors.

Do these growing islands slowly merge into a single, all-encompassing island that takes over the entire shoreline? To answer this question, think about what happens when two islands of opposite actions meet. At the boundary between these islands, we have two nearest neighbors taking different actions, but each one takes the same action as the neighbor on her other side. Thus, each of these nearest neighbors has one neighbor (the island mate) doing the same action and one neighbor (the boundary mate) doing the opposite. Given majority rule, neither one will want to change her chosen action. Thus, when two islands of opposite actions meet up, they both stop growing at the meeting point and a stable border is established.

Given the above, we now have enough insights to understand the dynamics of Lakeland. Whatever the random initial conditions, we will see islands of common action emerging from those spots on the shore where at least two nearest neighbors happen to take the same action.* Each of these islands will lock into having all of the island mates taking the identical action for the rest of the season, though that common action will vary across the different islands. Over time, residents that are not part of any existing island eventually get accreted into an island. When two islands of opposite action meet, a stable boundary is formed. These processes eventually lead Lakeland to a stable state that has contiguous groups of residents all taking the same action, with that action alternating as we go from group to group around the circle (see Figure 8.2).

Thus, Lakeland breaks down into a set of very stable groups pursuing very different actions, even though all of the residents follow the same behavioral rule. Moreover, the formation of these groups is tied to the initial conditions. If we rerun the model with new initial conditions,

---

* There is one perverse case that can occur when the initial conditions and number of agents are such that the actions perfectly alternate as we go around the circle. In this case, every resident will switch her action at each time step, and the system will never settle down. The likelihood of such a case arising is vanishingly small as the system increases in size.

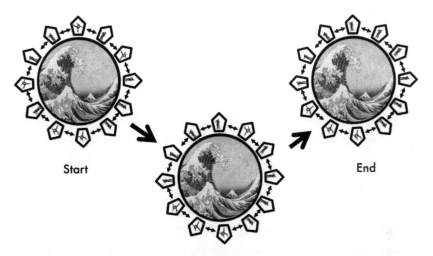

FIGURE 8.2: The community of Lakeland begins with random initial actions (left), evolves based on majority rule (middle), and finally reaches a stable configuration (right).

we might find that one season's meticulous lawn keeper becomes next season's cad letting her lawn go to seed.

Models become valuable when their insights can be applied to situations far beyond their initial motivation. So even if focusing on lawn care in Lakeland doesn't seem of interest in and of itself, there are in fact a number of phenomena, such as lawn care, home maintenance, and what color you paint the exterior of your house, that are similarly influenced by social behavior and that can affect everything from property values to the long-term stability of a neighborhood. Thus, a basic model of lawn care can give us insights into how neighborhoods can fall apart, and perhaps even suggest policies that might put them back together—such as strategically targeting particular residents for behavioral changes that will result in large positive impacts on the overall state of the system.

A variety of other social behaviors may be influenced by neighbors. Consider education. The desire to do your homework (rather than go to a party), participate in class discussions, or even go to college is often influenced by the actions of your friends, and thus a Lakeland-like model may offer insight. Similar forces may influence criminal behavior, as the actions of one's neighbors may encourage or discourage criminal activities, ranging from selling drugs to joining gangs. Indeed, in some communities, ignoring your lawn is viewed as an offense that is at best antisocial and perhaps even illegal.

Another obvious set of Lakeland-like models might involve religious and political choices. Religious practices, from celebrating particular holidays to decorating your house in lights to the choice of a religion itself, are often influenced by social networks and one's desire to conform. Similarly, views on political issues and choice of political party can be influenced by social networks.

In Lakeland, we assumed that everyone lived on a circle, and that social influences came only from one's nearest neighbors. This is a very extreme and sparse social network, and in more realistic models we might incorporate more complicated networks. For example, even in Lakeland, residents might be influenced not only by their nearest neighbors but also by their next-nearest neighbors. Furthermore, perhaps they can see across the lake, so the actions of more remote neighbors might be influential as well.

It has been found that changes to the structure of a network often have a big influence on system-wide behavior. Consider the problem of relaying a message to someone you don't know via people you do know. Suppose that you want to send a message to a randomly chosen person in the network and that you are only allowed to pass this message to someone you are directly connected to, who in turn must pass it to someone she is directly connected to, and so on, until the message arrives at its destination. What is the smallest number of links (on average) that it will take for you to make the needed connection?

In Lakeland, where everyone lives on a circle and is only connected to her immediate neighbors, a randomly chosen recipient is likely to be one-quarter of the way around the circle from the original sender in one direction or the other (at most, the recipient and sender can be directly opposite each other, which is halfway around, so on average they will be at the one-quarter mark). Since messages can flow only across links in the network, the most direct route to the target will have the sender passing the message to her nearest neighbor in the shortest direction to the target. The neighbor will do the same, and so on. Therefore, the message will be passed through, on average, a quarter of the population of Lakeland before it arrives at the target. Note that as the population gets larger, the length of time to get the message to the target increases linearly. If there are 6 billion people arrayed around the lake, it will take, on average, 1.5 billion steps to deliver the message.

In Lakeland, everyone knows only her two nearest neighbors. In real networks, while we likely have a lot of very local connections, we often have a few more distant ones as well. So let's modify Lakeland by giving some of the residents a connection to a randomly chosen person. This new network is like our original Lakeland, with everyone still connected to their nearest neighbors, but with the addition of a few new connections randomly spanning the lake. This new type of network (see Figure 8.3) is known as a small world network, for reasons that will become obvious in a moment.

Passing messages in a small world is very different from what we originally did in Lakeland. In Lakeland, we had the tedious process of going around the circle from nearest neighbor to nearest neighbor until we finally arrived at our target. In a small world, you can exploit the new, long-range connections to expedite delivery of the message. A small world resembles something akin to a network of local roads and highways. If you want to go somewhere fast, you take a few local roads to get on the highway, stay on the highway until you can exit near your destination, then proceed to your destination on the local roads.

While it is clear that small world networks should speed up message passing (after all, it can't take any more steps than before, since you can always revert to the outer-ring, nearest-neighbor approach if need be), it is surprising how

FIGURE 8.3: A small world network formed by taking Lakeland and adding some random, distant connections. The addition of such connections dramatically alters the message-passing dynamics of the system.

much less time it takes. Taking the example of 6 billion residents above, and assuming that each resident knows thirty people, then the expected number of passes is only about 6.6—it's a small world after all! Recall that for a Lakeland with 6 billion residents we needed 1.5 billion steps if each resident had only two friends. If we assume thirty nearest neighbors, the equivalent calculation would require a message in Lakeland to be passed 100 million times.

Thus, if we are willing to accept the assumptions of small world networks, there is a little more than six degrees of separation between you and someone else on the planet (if we allow for the loss of a billion or so folks given their inability to participate on various grounds). The small world model assumes that random connections are possible between any two people in the world, and this assumption may not hold, so consider the estimate of six degrees of separation as a lower bound. Regardless, the result is remarkable.

Researchers have investigated various networks, including coauthors of scientific papers, people who friend each other on Facebook, the links that make up our electric power grid, biological regulatory networks that control the expression of genes, the connections across neurons in simple brains, and links across web pages, to name just a few. The evidence is slowly accumulating that many of these networks have a deep common structure that may provide a basis for developing some unified theories of how such networks arise and behave.

In 1969 Thomas Schelling created an interesting model similar to the ones discussed above. Schelling was interested in understanding issues surrounding segregation. Instead of people arrayed around a lake, suppose that each resident occupies a square on a checkerboard (where not all of the squares are occupied). Each resident in the interior of the board is surrounded by eight neighboring squares.

Suppose that each resident is either a type X or O. We assume that the two types of residents are tolerant of each other and that as long as at least 30 percent of their neighbors are the same type as they are, they are content to stay in place. However, if the proportion of same-type neighbors drops below 30 percent, that resident will randomly relocate to one of the empty squares.

Given the very weak preference for having neighbors of the same type, one might expect that the world described by this model would quickly settle down to a state with very little segregation between the two types. Unfortunately, the actual behavior confounds such an expectation.

Figure 8.4 shows the arrangement of residents both randomly arrayed on the landscape (top) and after everyone who wants to move has done so (bottom). At the start of the model, since residents are randomly placed on the board, on average 50 percent of a resident's neighbors are of the same type and 50 percent are different. If you look at the initial configuration of residents, there is little evidence of segregation—whatever patterns you perceive are due to your mind wanting to put order and pattern on the randomness (this is a common phenomenon—for example, random sequences of coin flips look far more like *HTHH-HTTH* . . . than *HTHTHTHT* . . .).

From the initial starting conditions, we allow any resident who has 30 percent or fewer neighbors of the same type to randomly relocate. As can be seen in the figure, such a process quickly leads to large, segregated neighborhoods. Indeed, we find that after the system settles down, each resident, on average, has around 70 percent of her neighbors being of the same type. Thus, a slight preference for having

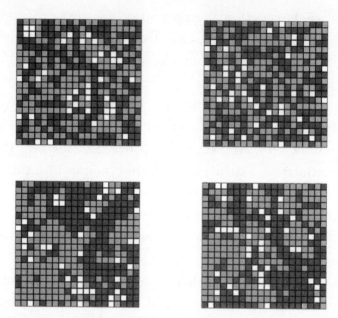

FIGURE 8.4: The Schelling segregation model with 360 agents, where residents move if 30 percent or fewer of their neighbors are of the same type. The two types of residents are shown by the differently shaded cells, with the white cells being unoccupied. Both random starting states (top) lead to the corresponding ending states directly below. In both runs of the model, what begins as a world where on average the likelihood of a similar-type neighbor is around 50 percent transforms into a segregated world with a greater than 70 percent likelihood of similar-type neighbors. (The program that generated this output was created by Robert Hanneman.)

at least 30 percent of your neighbors being like you leads to having 70 percent of your neighbors being like you.

You might at first think that the random mixing we used to initialize the system would be sufficient to keep everyone in place, as on average each resident has 50 percent of her neighbors being similar. Of course, the 50 percent is an overall average, and some residents will live in neighborhoods with a higher or lower percentage of similar residents. Thus, some of the randomly placed residents will find themselves in neighborhoods with an insufficient number of same-type neighbors, and they will move. When a resident moves, each of her eight neighbors loses a neighbor of that type, and this may be sufficient to tip the balance of same-type residents for some of the old neighbors, inducing them to move as well. As the proportion of a given type of resident in a neighborhood goes well above 30 percent, it not only becomes more stable to that particular type but also drives out the opposite type. Similar to what happened in Lakeland, stable configurations of contiguous, same-type-resident islands begin to form, and these slowly grow as they accrete any newly displaced, same-type residents that happen to land nearby.

We have seen before how positive feedback loops can cause a system to rapidly tip into a new, self-reinforcing configuration that is far away from its starting point. Schelling's system is governed by such feedback loops. Agents with a slight preference to be with same-type neighbors form positive feedback loops, with like begetting like.

If we alter the networks, we may induce very different behavior in the system. For example, the degree of segregation that arises in Schelling's checkerboard tends to increase with some reasonable alternative network configurations such as Lakeland's loops. In general, it can be shown that the key driving factor in these segregation systems is the

amount of overlap any given resident has with her neighbors' neighbors.

For the first part of human history, we were embedded in fairly static networks, consisting of some dense
connections across a small tribe with occasional, though
often transient, connections to outsiders. Over time, these
networks have grown far more dense and dynamic as we
have developed the ability to easily move and communicate across large distances. In the twentieth century, social
networks grew more connected as mass media developed
and a small group of people began to broadcast messages
to others.

More recently, with the advent of computers, our networks have become even more complex, as we become
"friends" with people we have never met in person who
live in locations that we have never visited. We now interact with anywhere from a few dedicated friends to thousands of followers through email, blogs, status updates, and
144-character messages. We find ourselves at the nexus of
overlapping networks consisting of large groups of friends,
coworkers, and various other contacts. We are only beginning to understand the impact of this new, hypernetworked
world in terms of complex social dynamics. Posting a picture of your freshly mowed lawn on Facebook may have
social impacts far beyond your immediate neighborhood.

NINE

From Heartbeats to City Size: *Scaling*

> "Villains!" I shrieked, "dissemble no
> more! I admit the deed!—tear up the
> planks! here, here!—it is the beating of
> his hideous heart!"
>
> —EDGAR ALLAN POE, "The Tell-Tale
> Heart"

MAMMALS LIVE, ON AVERAGE, FOR ROUGHLY 1 BILLION HEART-beats. No matter how large or small, their lives are ticking away with every beat. Thus, a mouse, with an average heart rate of about five hundred beats per minute, is expected to live for four years. A human, with fifty beats per minute, lives for around forty years. With a fixed number of lifetime heartbeats, the slower your base heart rate, the longer you live.

Such relationships are useful for predicting, and perhaps even understanding, the world around us. From a mouse to a blue whale, and every mammal in between, we can now make a useful prediction about an animal's life span knowing only its pulse rate. Moreover, heartbeat is tied to other physiological features, such as body mass and metabolic rate, so these too can be predicted. The existence of such scaling relationships suggests that there may be some greater, universal laws that underlie these systems.

Mathematically, there are many ways one variable could be related to another. Two variables could have a linear relationship, such as $y = x$. A specific type of relationship that arises regularly in a variety of systems is known as a power law. A power law states that something scales as something else raised to some fixed power, such as $y = x^2$. For example, the area of a square is equal to the length of its side raised to the power of 2 (that is, multiplied by itself twice). If we double the length of each side, the square doesn't double in area, it quadruples ($2 \times 2 = 4$). The volume of a cube scales as the length of its side raised to the power of 3 (multiplied by itself thrice). Therefore, doubling the length of a cube's sides results in eight times the volume ($2 \times 2 \times 2 = 8$).

These geometric relationships may seem too simple to shed much light on how complex systems work, but they have some interesting implications for biology. In a world of roughly cube-shaped animals, as we double their size their surface area (think amount of skin) goes up by a factor of four, while their volume (think guts) goes up by a factor of eight. Thus, there is less surface area per unit of volume, which makes it is easier to stay warm when you get bigger (since you lose heat through your skin and generate it via your guts). These geometric relationships also imply that as animals grow larger, their bone structure has to change in a disproportionate way, as the bones' ability to support the animal (which is tied to the bones' cross section or area) grows only as the square of size, while the weight of the animal (volume) grows as the cube. The good news of this latter result is that your typical B-movie attack of the giant whatever caused by a clumsy janitor knocking over a vat of radioactive whatnot during a late-night laboratory cleaning would be doomed from the start, as the creature's proportionately sized appendages would collapse under its disproportionate weight—elephants have thick legs for a reason.

Knowing the value of the power in a power law tells us how the system scales. If the power is 1, then as we double the independent variable (say, the length of a stick), we simply double the dependent one (say, the stick's weight). When the power is greater than 1, the system scales superlinearly, so when we double the independent variable, the dependent variable more than doubles. This is the type of scaling we saw in the area and volume relationships above, where the powers were 2 and 3, respectively. Finally, if the power is below 1, the system scales sublinearly, and doubling the independent variable results in less than a doubling of the dependent variable.

Allometry is the study of relationships between the physical and physiological features of organisms. Such studies date back at least to Otto Snell's work in 1892. In the 1930s, Max Kleiber noted that the metabolic rate of an animal scales to its mass raised to the ¾ power (that is, it scales sublinearly). The metabolic rate tells us the amount of energy needed for an organism to survive. A power of ¾ implies that we only need two times the energy to sustain two and a half times the mass. In general, this relationship implies that as animals get larger they are more efficient in the amount of energy needed per unit of mass.

Since metabolism is tied to all kinds of other factors, such as oxygen intake and heart rate, it is not surprising that scaling laws exists for these factors as well. Breathing and heart rates scale with mass to the −¼ power. Note that if we have a fixed number of breaths or heartbeats in a given life span, then this implies that life span scales with mass to the ¼ power. Under this type of scaling, if you are sixteen times as big, you will live twice as long.

With allometric scaling laws in hand, we are now able to predict some critical outcomes—such as metabolism and life span—knowing only an organism's mass (see Figure 9.1). While many of the examples above focused on mammals, you can extend the allometric scaling laws to other organisms as well. Even at extremely small scales, such as a single cell, they still hold. Thus, over a vast swath of life on earth encompassing more than twenty orders of magnitude of mass, we find a simple law that connects them all.

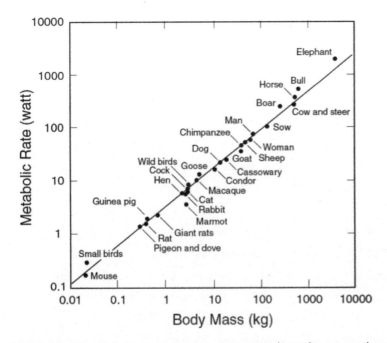

FIGURE 9.1: Metabolic scaling across various animals. Both axes use logarithmic scales, and therefore under a power law the plotted points should all fall on a line. Here the implied power is ¾, leading to sublinear scaling. This implies that as organisms get larger their relative metabolic needs decrease—an organism twice as large needs less than two times the total metabolism. *(Figure courtesy of Geoffrey West.)*

When such laws appear, it suggests some underlying mechanism driving the entire system. In the case of metabolism, such a mechanism has been identified by Geoffrey West, Jim Brown, and Brian Enquist. The idea behind this mechanism is that even complex structures such as bodies face some constraints. Here, the constraint is on physical limits to the size of features required to exchange nutrients, such as the capillaries in your circulatory system. If there is a limit to how small capillaries can become, then to grow bigger you need to find a way to pack enough of them into the larger space so that they can oxygenate the tissues. Such a requirement constrains the whole system.

Imagine a group of thirsty people sitting in a single row of a stadium on a hot day. Suppose we have a vendor on the aisle who can hand the first person in the row a cup, and that person will either drink it if she is thirsty or pass it on to the next person, who will follow the same behavior. Here, the constraints are the ability of the vendor to hand out new cups and the size of each cup. If very few people are in the row, the water will easily satisfy everyone's thirst. However, as we add more people to the row, if everyone drinks the same amount as she did before, the water will never reach the people located at the end of the row. But if everyone's thirst is reduced—that is, if we metaphorically lower each person's metabolism—then everyone's thirst can be satisfied. To sustain longer rows we need patrons who are less thirsty. We can even quantify how much less thirsty they must be: in the case of our stadium it is 1 over the number of patrons (giving us a power of -1 for thirst per person, and of 0 for total thirst). Thus, when we add an eighth person, if everyone's thirst goes down by one-eighth, the existing amount of water will still suffice. Moving from our one-dimensional stadium to three-dimensional biological systems is more involved and leads to a power of $-\frac{1}{4}$ for metabolism per unit of mass and a power of ¾ for total metabolism, but the fundamental idea is the same.

So we now have a simple biological relationship, in the form of a power law. It provides a nifty summary of the world around us, along with an identifiable reason for why such a law exists: physical constraints. Of course, the notion of "law" here is more along the lines of posted speed limits on a road (it is just a good idea) versus a fixed law such as gravity, as we do see some violations. For example, primates and parrots live about twice as long as one would predict using the scaling law. This may be tied to the longer developmental stages needed for their relatively bigger brains to realize their evolutionary potential. Humans are even greater outliers, likely because of improved sanitation and medicine (thus allowing us to live far longer than the expected forty years). Some domestic animals, such as dogs and cats and perhaps even cattle and horses, also overperform. This may be due to domestication and artificial selection. Thus—and oddly, given the scaling law—small dogs tend to live longer than large ones (though mice, guinea pigs, and rabbits tend to line up as expected, pet-buying parents beware). Animals that fly, including both birds and bats, tend to live longer than expected. This might not be too surprising if flying somehow implied lower metabolism, but the opposite is true, as beating wings require more heartbeats, not fewer.

While it would be nice if the "law" worked perfectly, even an imperfect law is useful. In science we often face trade-offs between having a complete understanding of some specific thing and an incomplete understanding of many things. In biology, for example, a huge amount of effort has been spent trying to understand single organisms, and there are biologists who specialize in a particular species of worm (and, within that species, specific aspects therein). While such studies provide key insights, Murray Gell-Mann's "crude look at the whole" suggests that there is power in being able to develop generalized insights across broad domains, even if these efforts sometimes fail. Indeed, such failures often provide new insights. Thus, knowing that big-brained or domesticated creatures tend not to follow the general scaling law may give us some new insights if we are careful not to tell just-so stories.

Finding a general law in one area may inspire us to search for related laws in other domains. If biological systems face constraints leading to scaling laws, then perhaps other systems do so as well.

Lewis Fry Richardson pioneered modern techniques for forecasting both the weather and wars. In his *Statistics of Deadly Quarrels* (1950) he gave war a statistical face. Table 9.1 shows some of his key data. As can be seen, the higher the number of deaths in a given war, the fewer such wars we observe (thankfully). The numbers in parentheses provide some useful approximations to the data, and using these, we can see that as deaths increase by a factor of ten, the number of wars decreases by a factor of three.

Approximate Deaths	Number of Wars
10,000,000 (10^7)	2 (2×3^0)
1,000,000 (10^6)	5 (2×3^1)
100,000 (10^5)	24 (2×3^2)
10,000 (10^4)	63 (2×3^3)
1,000 (10^3)	188 (2×3^4)

TABLE 9.1: Deaths in warfare, 1820–1945, from Richardson (1950). The values in parentheses provide useful approximations to the data.

Richardson's findings can be translated into the language of power laws. Doing so, we find that the number of wars is proportionate to the number of deaths raised to roughly the $-\frac{1}{2}$ power. This implies that when you double the number of deaths, the expected number of wars is 70 percent of the previous value. One implication of this relationship is that we should expect one war with around 42 million deaths at some point. Of course, the power law doesn't tell us when such a war will happen, only that one such war is expected if the predicted distribution holds.

George Kingsley Zipf was an American linguist interested in the statistics underlying word use in languages. It's not too surprising that if we count the number of occurrences of various words in a text, we find that some words are used far more often than others. The frequency with which a word is used (designated by its rank) is described by a power law with an exponent of -1. So the word that is the second most commonly used in a text will occur about half as often as the word that's most commonly used. The third-most-common word occurs one-third as often, and so on. This relationship holds across a variety of languages (including languages that are randomly generated).

Zipf-like laws occur in other contexts as well. For example, the distribution of the size of cities or corporations also follows Zipf's law. The largest city in a country has about twice the population of the second-largest, three times that of the third-largest, and so on. As with Zipf's law of languages, this relationship holds across a variety of different contexts (such as different countries or time periods), so, as before, there appears to be a certain universality in these observations.

Scaling laws may provide some useful insights into our ability to survive as a species. World population is roughly 7 billion people, and it is growing at about 1 percent per year (which implies that it will double in size every seventy years). Throughout human history, the majority of people lived in rural areas, though the proportion of people living in urban areas has been steadily increasing. Just recently the balance has shifted enough so that the majority of the world's population now lives in cities.

Cities are not all that different from biological organisms. They have metabolisms tied to the flow of energy and people along various transportation and communication networks that produce knowledge and economic outputs along with various streams of waste that get processed and released into the surrounding air, water, and land. So it may not be too far-fetched to think that universalities similar to what we see in biological systems might also apply to human-made ones, and if that is the case, urban systems may be governed by related scaling laws. If such laws exist, they may give us some insights into what the future holds for humankind.

Luis Bettencourt, José Lobo, Deborah Strumsky, Geoffrey West, and colleagues have calculated power law coefficients for a variety of urban metrics tied to a city's population size. Some metrics, such as the amount of road surface or gasoline sales, scale sublinearly, implying that as the population of a city gets larger, each person uses less of that resource. That is, bigger cities tend to have lower gasoline sales and less road surface per capita than smaller ones. Intuitively, this makes sense, as cities tend to build up rather than out, and that requires fewer roads, makes public transportation more viable, and leads to more energy-efficient transportation overall—in general, larger cities tend to economize on such infrastructure. There are other metrics, such as economic output, inventive activity (measured by, say, patents or R&D employment), crime, and disease, that scale superlinearly. Thus, larger cities are relatively more economically productive and creative than smaller cities, along with being more crime- and disease-ridden. This superlinearity tends to be tied to the more social elements of cities. Finally, a number of metrics, mostly linked to individual human needs such as housing, consumption of household resources, and employment, scale linearly, implying that on a per capita basis, all cities are the same.

These power law coefficients are preliminary estimates based on existing data. Moreover, they provide only a snapshot of the situation, and as we move to cities of vastly different size from those we have now, or as new inventions alter our opportunities, we might see these laws diverge as limits to growth start to bind or as new technological life rafts are deployed. Nonetheless, they do provide a sense of our future.

If the estimates are to be believed, as the world's population grows, concentrating more and more people in urban areas, megacities will relieve some of the demands for resources such as roads and fuel. Unfortunately, increasing urbanization will not mitigate the demands for individual needs, such as housing and electricity, which rise linearly.

It's the superlinear factors that likely hold the key to our future. The old woes of crime and disease—prominently featured in the dystopian views of most science fiction movies—will likely increase per capita as cities become bigger. Balancing this unfortunate scaling of woes is the prospect of per capita increases in economic growth and inventiveness in the emerging megacities.

Like the steady beat of a heart, world population continues to grow and concentrate into urban centers. Perhaps out of such concentrations, an inventive spark will emerge that allows us to prolong our existence beyond the allotted number of beats.

TEN

From Water Temples to Evolving Machines: *Cooperation*

> Now join your hands, and with your
> hands your hearts.
>
> —WILLIAM SHAKESPEARE, *Henry VI*

B ALINESE FARMERS HAVE GROWN RICE ON TERRACED HILLSIDES FOR the last few centuries (see Figure 10.1). Rice demands water, as several important biochemical cycles that govern the production of the rice paddy ecosystem, such as soil pH, temperature, nutrient circulation, aerobic conditions, and microorganism growth, are directly tied to the carefully controlled flow of water. Thus, accompanying the terraces is an elaborate, gravity-fed irrigation system dependent on seasonal rivers, groundwater flows, and the creation and maintenance of various irrigation canals, tunnels, and diversion weirs. While these irrigation works are

FIGURE 10.1: Rice terraces in Subak Pakudui, Bali, currently flooded to control pests. These terraces are a striking modification of the landscape, representing both a coherent synthesis and an abject alteration of nature's inclinations. *(Photograph courtesy of J. Stephen Lansing.)*

impressive—especially considering the difficulty of digging kilometer-long tunnels and the like using hand tools and primitive surveying instruments—they cannot overcome the inherent scarcity of water.

Given the scarce water and the lack of central control of Bali's farmers, economists would expect to see a harsh, competitive outcome, best characterized by Thomas Hobbes in *Leviathan*: "And the life of man, solitary, poore, nasty, brutish, and short." This prediction is due to the presence of externalities—costs or benefits imposed on parties outside of (external to) the immediate transaction—within the system. In a world with scarce, gravity-fed water supplies, we might expect that upstream farmers would pay no heed to the needs of downstream farmers. Thus, an upstream farmer, thinking only of her own good, is willing to consume additional water as long as it results in at least a trivial gain in her own output, even when passing that water downstream might allow other farmers to reap a much larger gain. In such a situation, we could reallocate the water and maximize the total crop—in theory we could give every farmer as much crop as she was getting before and still have some left over, which could make at least one person better off.

Externalities represent just one example of where individual incentives result in inferior outcomes. Such situations, where being competitive makes you slightly better off while being cooperative makes you remarkably better off, are all too common.

Notwithstanding the Hobbesian prediction for Balinese rice farming, thanks to the work of Steve Lansing and his collaborators, we find on the island an example of system-wide cooperation among the farmers leading to many centuries of sustainable agriculture. Rather than hogging the water, the upstream farmers carefully coordinate and cooperate with the downstream ones, resulting in much larger overall harvests.

One potential clue to this cooperative mystery is that we find an elaborate religious system of water temples in Bali that closely parallels the physical system of terraces and irrigation works. Individual weirs in the system are associated with shrines, and those shrines get aggregated into temples dedicated to agricultural deities. Thus, local weirs are nested into local temples that get further nested into still other temples, with the various aggregations of these pieces being closely associated with the underlying irrigation system and the physical watershed. The congregations of the shrines and temples meet once a year to coordinate each individual farmer's use of water.

While it's tempting to end our story here, knowing that a religious institution has arisen to solve the externality problem and has resulted in harmony and happiness throughout the system, such a conclusion is far less interesting than what really drives this system's cooperation.

Like all agricultural systems, the rice ecosystem must overcome attacks by pests, including insects, rodents, microorganisms, and the diseases they bear. Pests can sometimes destroy almost the entire crop. The ultimate amount of pest damage is tied to both natural and human factors, such as patterns of water flow and harvesting.

As crops grow, so do the pests. When a crop is harvested, the nutrients are removed, and pest populations crash. However, if the newly fallow field is near an unharvested field, the pests will move over and continue to grow. This latter piece of ecosystem dynamics results in the second major externality found in the Balinese system—a farmer harvesting her rice crop may impose an uncompensated cost on neighboring farmers as the pests from her crop migrate to the neighboring crops.

In the 1970s, the Indonesian government inadvertently tested the ecosystem dynamics of Bali's rice fields. Based on advice from consultants at the Asian Development Bank, the Indonesian government undertook a massive redirection of agricultural policy and legally mandated the double- and triple-season cropping of newly developed, high-yielding varieties of rice. These mandates led to the abandonment of the temple system (noted in official reports as a Balinese "rice cult") of coordinated, cooperative agriculture.

Soon after this change, reports of "chaos in the water scheduling" and "explosions of pest populations" began to trickle into district agricultural offices. The pest problems that arose were first mitigated by the introduction of new crop varieties resistant to the pests at hand. Alas, nature finds a way, and these new varieties soon succumbed to new pests. Government reports begin to read like a tragic farce: the plague of brown planthoppers was reduced by the introduction of the planthopper-resistant rice strain IR-36, but this new variety of rice was quickly overwhelmed by tungro virus, which was countered by the introduction of PB-50, which unfortunately was susceptible to brown leaf spots caused by the *H. oryzae* pathogen, and so on. Crop losses due to pests approached 100 percent during this time, and Balinese farmers remember this period as a time of *poso* (hunger and harvest failures).

The discussion above contains the needed clues for piecing together the emergence of cooperation in Balinese rice farming. As the irrigation systems developed, the upstream farmers likely ignored the needs of the downstream ones and took whatever water they wanted. Rice

requires periodic floods, so presumably the downstream
farmers were able to exist only by offsetting their planting
so that their peak use of water occurred during ebbs in
the water demands of the upstream farmers. This resulted
in staggered harvests, where the upstream farmer would
harvest her crop while the downstream farmer's crop was
still growing and vice versa. As long as the pest populations
were modest and the fields far apart, this was a workable
system. As the population grew, however, the demand for
rice increased and more terraces were developed and put
under production. This resulted in more closely packed
fields and a monocultural ecosystem, two factors that favor the growth of pests and their transmission across fields
if farmers do not coordinate fallow periods.

Since the upstream farmers have first access to the water, they are best off if they take as much water as they
want *and* minimize damage from pests by having the same
fallow periods as the downstream farmers. However, if the
externality costs of the pests flowing between the fields
are low relative to the damage caused by scarce water, the
downstream farmers would rather wait for the bigger water
flows and not have identical fallow periods. Under such
conditions, the farmers find themselves in a strategic situation akin to a duel between a baseball pitcher and a batter,
where the pitcher wants to throw the ball where the batter
isn't swinging.

As the damage from pests increases, however, a remarkable system-wide transition becomes possible. If the

external costs from pests exceed those from scarce water, downstream farmers will want to cooperate and plant at the same time as the upstream farmers, since now it is better to lose due to water scarcity than to pest damage. When this happens, a new equilibrium becomes possible in which the farmers coordinate so that both fields simultaneously lie fallow, killing off the pests. Under certain conditions, increasing the amount of damage caused by pests can, rather counterintuitively, *increase* the total output of crops. This occurs because as pests become worse, they cause the system to transition into a cooperative regime under which the farmers coordinate planting, and since both farmers pay a cost (now avoided) from pests while only the downstream farmer pays a cost from water scarcity, the total production in the system increases.

So why do we see temples if the motivation for cooperation is so directly tied to the farmers' costs and benefits? To cooperate, the farmers need to coordinate. Thus, there is a role for some type of institution—the water temples—to serve as a coordination device. Since all farmers want to coordinate, it is in their self-interest to seek and follow whatever advice is given by the temples. Even without threats of force, fear of calamity, or ostracism, the water temples have an implicit power to dictate planting times to all of the farmers (see Figure 10.2).

The story of farming in Bali is a tale of cooperation arising in a complex system. We began with what, on the face of it, should be a disastrous situation in which upstream

FIGURE 10.2: Offerings to the Goddess of the Lake during the festival of the full moon of the tenth Icaka month on the island of Bali, March 2011. This temple controls the access of various irrigation systems to a key water reservoir in the ecosystem. *(Photograph courtesy of J. Stephen Lansing.)*

farmers, paying attention to only their own welfare, hog all the water, resulting in a greatly diminished output of crops. Then we added a complex set of dynamics, both natural and human, that realigned incentives in such a way that cooperation became possible and crop output increased, making everyone better off. To realize this new outcome, there was a need for a coordination device, and thus a new social niche opened up for a religious institution that, rather than practicing some arbitrary ideology, was in fact driven by the underlying hydrology, crop growth patterns, and pest population dynamics, regardless of whether any of its founders or practitioners actually realize it. The interaction of complex natural and social systems led the entire system to a remarkably better place than would otherwise arise.

IN THE STRUGGLE FOR SURVIVAL, COOPERATION IS ONE strategy that tends to provide a definitive edge. "Tho' Nature," as Tennyson says, may be "red in tooth and claw," the ability to cooperate rather than compete often allows a group to thrive far beyond its apparent means. Cooperation leverages fitness, and examples of this abound across all levels of existence. A bacterium can do little harm to a host, yet a group of bacteria, coordinating their attack through a set of chemical signals, is deadly. A small fish is an easy target for a predator, while a school of such fish can move with relative impunity. Humans in groups, whether undertaking trade in a village or fighting within an army, are far more likely to survive than their solitary brethren.

Thus, understanding how cooperation can emerge and be maintained is a key issue in furthering our understanding of how interacting agents survive in complex worlds. The case of Balinese rice farming illustrates one approach. To understand the embedded complexities of that system, it took a variety of contributions from across the sciences. Anthropologists looked at current farming and religious practices on the island and, in conjunction with archaeologists, reconstructed past practices. Historians investigated the impact of the green revolution on Balinese agricultural policy. Biologists, agricultural specialists, hydrologists, and geographers developed insights about the ecosystem, delineating all of the interactions among crops, water, and pests. Computer scientists developed agent-based models of farmers making adaptive decisions about cropping choices. An anthropologist and I used ideas from game theory to develop the sparse model of choice between upstream and downstream farmers used above. Together, we were able to put together the various pieces of the puzzle and develop a coherent story of the emergence and maintenance of co-operation on Bali.

Of course, these insights required an amazing span of professional expertise and effort. An alternative approach to understanding cooperation is to rely on a relatively stark abstract model. Such an approach is perhaps the polar opposite of what my colleagues and I undertook on Bali. But if we are skilled, and lucky, the details that we ignore will not matter much, and we can use this abstract model to gain some new general insights.

The quintessential, stylized problem for exploring co-operation is known as the prisoner's dilemma. In its original version, two co-conspirators have just been apprehended by the police and placed in separate cells. Although the police suspect them of committing a capital crime, there is little evidence, so if neither prisoner confesses, both will be jailed for one year. The police offer each prisoner the following deal: if he confesses and becomes a witness for the state, then he can go free, though his accomplice will be put to death. The only proviso to this deal is that if both prisoners confess, each will be jailed for ten years. Each prisoner must decide what to do without any knowledge of whether the co-conspirator has confessed.

Each prisoner faces an interesting dilemma. If he confesses and his accomplice stays quiet, then he will go free rather than spending a year in jail. Similarly, if he confesses and his accomplice also confesses, then he will only get ten years in jail instead of being put to death. Regardless of what the accomplice does, the prisoner is always better off confessing, that is, defecting on his accomplice. Of course, both prisoners face the same situation, so for both of them it is in their interest to confess, which means that they both will spend ten years in jail. If, instead, they could have cooperated and stayed quiet, they would have been jailed for only a year. As we saw before, while being competitive makes you slightly better off, being cooperative makes you remarkably better off.

If the prisoner's dilemma were only about prisoners, it would be of passing interest. However, the basic framework here captures many interesting scenarios. Opposing soldiers dug into trenches on a battlefield can choose to make a predictable and passive show of force that both allows them to stay safe and keeps their commanding officers at bay (that is, they can cooperate), or each side can attack the other (that is, they can defect). Infantryman or lionesses on the hunt can stay at the front of the line (cooperate) or fall back a bit to let others bear the brunt of the attack (defect).

Two rival firms can tacitly keep their prices high (cooperate) or provide hidden discounts to customers (defect). Fishermen can limit their catches to maintain a reproductive stock of fish (cooperate) or can violate such limits when others are not looking (defect). Polluters can limit their output of carbon dioxide (cooperate) or not (defect). Bacteria can release a toxin simultaneously (cooperate) or avoid doing so and save on energy (defect). Sellers on eBay can accurately describe their items and follow through on sales (cooperate) or be misleading before or after the sale (defect). And on and on.

Note that the labels "cooperate" and "defect" capture only the actions and respective incentives of the individual agents. They do not reflect the goals of society. In some cases, such as the rival firms setting either high or low prices, cooperation leading to a lack of competition and high prices is a bad social outcome. In other cases, such as the fisheries, having fishermen cooperate and limit their catch in order to ensure that the stock of fish can be maintained is a good social outcome. Regardless, in the absence of mitigating factors, the logic of the prisoner's dilemma leads to defection—which is a good outcome if we are consumers of a product such as oil and a bad one if we care about our planet's fisheries and carbon dioxide level.

Defection is the obvious outcome to predict if we are in the stark prisoner's dilemma world described above. However, if we alter some of the underlying conditions, cooperation may become a more reasonable outcome. For example, the two prisoners were not allowed to communicate with each other, yet if they could communicate (and trust each other's word), they would agree to stay quiet. Another feature of the game that encourages defection is its one-shot nature. When the players interact only a single time, the logic of "regardless of what the other player does, I'll be better off defecting" holds. However, if the players are going to play the game repeatedly, then the shadow of the future becomes important and cooperation becomes an attractive choice, since the short-term gain of defecting can be overwhelmed by the potential for a longer stream of cooperative benefits.

Cooperation in the prisoner's dilemma has been studied in a variety of ways. One method is to do case studies of real-world examples, such as trench warfare during World War I or the behavior of lobster fishermen in Maine. Another approach uses the mathematical tools of game theory to explore the limits of cooperation in a very abstract world. A third approach studies the game using experiments, either in the laboratory or in the field, where we observe subjects (ranging from college undergraduates to bacteria in a petri dish) under randomized and controlled conditions. All of these techniques have yielded useful insights into the origins of cooperation.

The different approaches all have strengths and weaknesses. Case studies provide direct evidence of cooperation arising in real-world contexts, though it is often hard to gather the needed historical data and analyze it given the complexities involved in the actual events. Mathematical approaches provide a nice abstract formulation of the problem, but at times such abstractions can be too stark. For example, a repeated game that is certain to end after a thousand rounds results in a dramatically different outcome from one that is only *expected* to end after a thousand rounds—in the former case the known final round leads to a defection on that round, which eventually unravels through all of the previous rounds, leading to mutual defection throughout the game, while in the latter case the uncertainty of the final round makes cooperative strategies possible. Experiments have been extremely useful in giving us some sense of the conditions needed to establish and maintain cooperation, though we are often limited by the subjects we can use, by our ability to uncover their underlying strategies, and by difficulties in creating suitable experimental environments.

To further our understanding of cooperation, we need to create new approaches that will allow us to gain better insights into the problem of cooperation. Over the last two decades, colleagues and I have been developing a high-technology hybrid of what has been done before. It has provided an important new window into the emergence of cooperation in complex systems.

The idea behind this new approach is to create an artificial world inside a computer. As with case studies, we want to allow for a fairly complex set of interactions among agents that can actively adapt to their experiences. As with the mathematical approach, we want to rely on some core, tractable, abstract concepts that can provide the needed structure and direction. As with the experimental approach, we want to be able to carefully observe and manipulate the world in which our agents interact. In the end, the approach combines key ideas from computer science, game theory, mathematics, and biology to give us a new view of cooperation.

Each of the various elements of this new approach is simple. We use the repeated prisoner's dilemma game as the foundational physics of the system. Agents in our artificial world play against one another by submitting a sequence of either cooperate or defect actions based on what happened in the previous periods, and then receive payoffs given by the game (here, instead of prison sentences, players receive appropriately scaled points).

Each player's strategy is represented by a simple computing machine called a finite automaton. These machines are composed of a set of internal states, with each state dictating an action to be taken and a set of transitions to other states that depend on the observed action of the opponent. Figure 10.3 illustrates one such automaton. While automata are simple in structure, they can lead to complex strategies that react to their opponent's actions based on conditional behavior, history, counting, and even randomness.

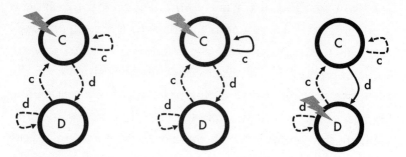

FIGURE 10.3: A simple two-state automaton. The leftmost figure shows an automaton with its two states represented by the labeled circles. Contained within each circle is the action (either cooperate (C) or defect (D)) that the automaton will take when it enters that respective state. The dashed arrows emerging from each state point to the transition state based on the observed action of the opponent in the previous period (either cooperate (c) or defect (d)—note the use of lower-case letters here). If this automaton always begins in its top state, it will initially cooperate (indicated by the lightning bolt). If the opponent also cooperates, the machine will follow the appropriate transition arrow (middle figure, solid arrow) and stay in its top state, so it will cooperate again. If the opponent defects (rightmost figure, solid arrow), the automaton transitions to its bottom state and defects next time. The logic governing transitions in the bottom state is similar to that for the top state. Thus, this automaton begins by cooperating and then mimics the opponent's previous move, a useful strategy known formally as tit-for-tat.

Finally, to allow the agents to adapt and improve their strategic machines, we use a simple version of artificial evolution called a genetic algorithm. As with evolution in the real world, a genetic algorithm selects machines for reproduction based on how well they are performing in the game, with the better-performing machines being more likely to be reproduced. Each machine's behavior is captured in the computer-coded description of its automaton, and like DNA, this description is passed on to the offspring with perhaps a few slight changes (mutations) that might either lead to a subtle change in the player's behavior or perhaps create one of Goldschmidt's "hopeful monsters" with a heretofore unseen and deviously clever strategy. As Darwin reminds us, "From so simple a beginning endless forms most beautiful and most wonderful have been, and are being, evolved."

While each of the elements of this computational model is simple, the artificial world that the model creates is not. We begin by randomly generating the machines. In such a world there is little order, as each machine cycles through its various states emitting random sequences of cooperate and defect. In this environment, machines that defect more than normal, due to the stochastic luck bestowed upon them when the strategies were randomly endowed, will do better than average, because in the absence of any order on the part of the opponents, defection leads to higher payoffs. Consequently, the initial evolution of this system favors machines that defect more often, and out of the chaos of our randomly created soup of life comes a wave of order red in tooth and claw, as defection takes over our computational ecosystem.

The machines that arise during this initial wave of evolution are surprisingly well structured. While the automaton has access to a large number of potential states, the surviving machines use very few of these, favoring simple structures that always defect. Even though larger machines could also implement an always-defect strategy, such machines are far more sensitive than smaller ones to mutations during reproduction. Given that most mutations are detrimental (here leading to cooperation when facing a sea of defection), smaller machines prevail. Thus, evolution initially creates a world composed of simple machines that always defect.

Imagine trying to gain a foothold in such a world. The only way to do better than the other players is to somehow establish cooperation with someone. However, always cooperating would make you much worse off, as cooperating against a defector gives you the lowest payoff possible (the death penalty in the example of our prisoners) and your opponent the highest (going free). So if a strategy spontaneously arises that always cooperates, it will be an easy mark for the defectors and will quickly die out.

Suppose, however, that two such cooperative strategies arise simultaneously. In this case, when they meet, they achieve a payoff much higher than average. Unfortunately, even this cooperative bounty will be insufficient to offset the losses incurred when these cooperative machines play the much larger number of defectors. Even if small numbers of unconditional cooperators arise, they will eventually be overwhelmed by defectors. This, then, is also not a viable way for cooperation to arise in this world.

There is a slightly different path that could allow for the emergence of cooperation. Suppose that agents could interact with only a select few of the other agents. If, say, cooperators can stick together and only play the game among themselves while avoiding playing the defectors, then they would receive very high payoffs relative to the rest of the world. Unfortunately, our model provides no explicit way to directly recognize one's opponents, as there are no externally observable features that would allow a machine to make some inference or categorization of its next opponent. Even if this were possible, the machines have no memory of past opponents. Thus, interacting only with cooperators is not going to work.

However, something akin to selective interaction— albeit a bit more clever—does allow an enlightened, cooperative path to arise in the model. While machines cannot recognize their opponent at the outset, the sequence of actions that is played as the game progresses may allow machines to recognize one another. Our initial wave of evolution resulted in a world where opponents always defect, and thus a machine could signal that it is different by initially cooperating. We saw that a blind strategy of always cooperating is doomed to failure, so the only way such a strategy could work is if the machine alters its behavior based on how the opponent reacts to the cooperative overture. If a machine cooperates and finds that its opponent does not reciprocate, then the machine can start to defect and avoid being exploited further. If, instead, the cooperative moves cause the opponent to alter its behavior and begin to cooperate, then the two machines can establish mutual cooperation and do well together. To establish cooperation, a machine needs to arise that is willing to take the short-term risk of cooperating (in the face of a world of defectors), given the potential long-term benefit of identifying and establishing cooperation with a like-minded machine. It also needs to avoid being exploited by an opponent that is unwilling to establish cooperation—that is, it must learn to cautiously cooperate.

Such a machine embraces the seemingly infeasible ideal of a strategy that plays only with cooperators and (indirectly) avoids defectors. While machines cannot explicitly avoid playing defectors, they can do so implicitly by recognizing the opponent's type during the initial play of the game. If an opponent is identified as cooperative, the machines can establish and maintain mutual cooperation for the remainder of the game. If an opponent is identified as a defector and one cannot avoid playing them outright, mutually defecting for the remainder of the game is a second-best solution.

One particularly fascinating aspect of the above scenario is that these new machines are spontaneously learning how to communicate with one another. Here, the initial actions in a game are also serving as a communication device that either signals cooperative intent or not. Thus, the evolving machines hijack their initial actions and repurpose them to serve as communication signals. This entails a short-term cost of taking less-than-ideal actions against some opponents, in the hope of achieving the long-term benefit of establishing cooperation.

Thus even in a world of all defectors, cooperation can emerge. If at least two cautiously cooperative strategies emerge simultaneously, they can receive higher-than-average payoffs, and evolution will favor their perpetuation.

While it is easy to see how a few cautiously cooperative machines could thrive in a nasty world of defectors and eventually take it over, that leaves the question of how such strategies can spontaneously arise in the first place. Cautiously cooperative machines embody a fairly sophisticated strategy that must first send some sort of cooperative signal, and then, based on the opponent's response, play appropriately and either establish cooperation or avoid exploitation. Such a strategy requires some careful coordination.

One way this coordination could arise is if machines simultaneously receive a set of mutations that reconfigure them into cautiously cooperative strategies. Unfortunately, the likelihood of having such a set of carefully aligned mutations is small—using the creationist-embraced analogy employed by the astronomer Fred Hoyle, it's like the chance of a tornado sweeping through a junkyard and assembling a functional Boeing 747. The alternative is that a single mutation somehow results in the needed strategy. While on the face of it this idea seems equally implausible, in reality it is not. If we start with a simple always-defect machine, a single mutation can have one of two possible effects. The first is that it alters the single action of the machine from defect to cooperate, turning it into an always-cooperate machine and thereby sounding its death knell. The other possibility is that the mutation transitions the machine into a heretofore unused part of the automaton, resulting in a radically different strategy.

By definition, unused parts of the machines are not tested against nature. Therefore, mutations that occur in these areas are not subject to evolutionary pressure, and the unused structure can drift around without any immediate impact on the machine's overall performance. Such alterations are known as neutral mutations, since the changes they make have no observable consequences on how the machine behaves, and thus no impact on the machine's immediate fitness. So even in a world of simple, always-defect machines, all is not static, as the unused parts of these machines undergo neutral drift.

With neutral drift, it is possible for a single mutation to result in a radical change in behavior, such as having an always-defect strategy become a cautiously cooperative one. Once we have a couple of cautiously cooperative strategies, evolutionary forces will be sufficient to tip the system from a world where everyone defects to one filled with cooperation. So the real issue for the emergence of cooperation here is how likely is it that two or more cautiously cooperative strategies will spontaneously arise.

One way for this to happen is that the needed mutation simultaneously occurs in two machines. This is possible, since evolutionary selection and reproduction can, at times, result in the neutral parts of a machine getting replicated across the population for short periods of time. If this happens, and if the neutral configuration is right, a single mutation at the same spot on two separate machines can lead to the creation of two cautiously cooperative strategies.

There are also conditions in which even a single machine becoming cautiously cooperative is sufficient to tip the system. If a single such machine arises, it will do slightly worse than those machines that always defect. But if the degradation in performance is not too extreme, the machine may survive and replicate, resulting in enough cautiously cooperative machines in the next generation to tip the system. Another possibility is that a lone cautiously cooperative machine, since it is facing opponents that have only ever known defection, will inadvertently trigger cooperative behavior in some of the always-defecting machines. The always-defecting machines have never encountered a cooperative action and don't, in an evolutionary sense, know what to do. Like dodo birds meeting sailors for the first time, one or more subservient always-defect machines can provide enough of a fitness boon to the lone cautiously cooperative machine to allow it to replicate in the next generation and, eventually, tip the entire system into cooperation.

Since evolution is always in search of weaknesses, the newly cooperative strategies must always remain vigilant and able to react to an opponent's defection, even after cooperation has been established worldwide. If not, then there is the possibility of a mimic arising that sends all of the right handshake signals to establish cooperation but then defects. To maintain such vigilance, machines need at least two active states. Tit-for-tat, shown in Figure 10.3, is sufficient to establish cooperation with similar-type machines but, at the same time, avoids being badly exploited by an opponent that always or occasionally defects. As always, there is evolutionary pressure on the more advanced cooperative machines to create structures that can withstand deleterious mutations that might cause the machines to malfunction.

Of course, the same forces that allow the emergence of cooperation can also conspire to destroy it. A population of cooperators can become evolutionary lazy if they are rarely tested by defection. If this happens, the strategies can drift to the point where the machines simply cooperate, either from the start of the game or after confirming the initial handshake. Once this happens, machines that always defect (in the first case) or mimic the handshake and then always defect (in the second case) can enter and take over the world.

Our cautiously cooperative strategies are, in essence, evolving the ability to distinguish self from other. If an opponent gives the proper handshake, then it is considered self. If not, it is other. Thus, cooperation emerges in this system by having strategies play against themselves, which easily solves the cooperative dilemma. This new route to cooperation is an interesting variant on kin selection, whereby cooperation emerges in a system because the agents share a common genetic basis. Here, the notion of kin spontaneously arises as a function of the handshake that provides an alternative sense of group cohesion. The notion that communication allows such cohesion is a tempting hypothesis. It suggests that the emergence of communication could be a key path to cooperation in social systems and ultimately to survival.

The observation that being competitive makes you slightly better off while being cooperative makes you remarkably better off may be a fundamental property of social worlds. Unfortunately, another fundamental property of these worlds is that individual incentives tend to favor competition over cooperation. That being said, our exploration of two dramatically different systems gives us a ray of hope. By focusing the various lenses used in the study of complex systems, ranging from careful anthropological work on the religious practices of Balinese rice farmers to the analysis of computational ecosystems driven by artificial evolution and abstract theories of automata, we find that cooperation can arise and be maintained, even in systems that seemingly favor competition.

Perhaps the joining of hands and hearts is easier than we imagine.

ELEVEN

From Stones to Sand:
Self-Organized Criticality

> Nothing is built on stone; all is built in
> sand. But we must build as if the sand
> were stone.
>
> —JORGE LUIS BORGES

HERE WE BEGIN WITH A PILE OF SAND AND, ALAS, END WITH ONE. We often observe that complex systems, after having developed a beautiful and seemingly robust structure, can collapse in an instant. Consider your body, a collection of billions of cells, each interacting and forming a recognizable and vital you. Yet all of those interactions, all that you are and could be, can cease in only a few minutes if, say, you experience a misplaced shock to your heart. Or consider a civilization such as the Maya in their Classic period, in which a vibrant Mesoamerican culture suddenly falls apart.

Is there something innate about complex systems that demands an inescapable vulnerability to collapse?

To explore this question, let's randomly sprinkle grains of sand on top of an empty table. At first, as the grains fall, they stay where they land. With time, an occasional grain lands on top of another grain, and as long as the new height is not much higher than that of the surrounding grains, it will balance. As the sand continues to pile up, we eventually reach a point where a grain can no longer balance where it falls, and a little avalanche ensues as the grain tumbles onto its neighbor. With few grains on the table, such tumbles result in a slight displacement of the ever-growing pile. However, as the sand continues to mound on the table, tumbles begin to cause imbalances at neighboring locations, resulting in new tumbles and a larger avalanche, perhaps even to the point where some sand falls off the edge of the table.

How such sand piles behave forms the core of a model of self-organized criticality developed by the physicist Per Bak. Sometimes a falling grain has little impact other than to add itself to the spot where it landed. At other times the grain begins an avalanche that triggers a chain reaction of additional grains tumbling across the pile. Indeed, avalanches of all possible sizes, following a well-specified probability distribution (yet another power law), characterize this system's behavior (see Figure 11.1).

At any given time during our sand-pouring experiment, we could pause and take stock of the conditions at any location on the table. At every location, the pile is

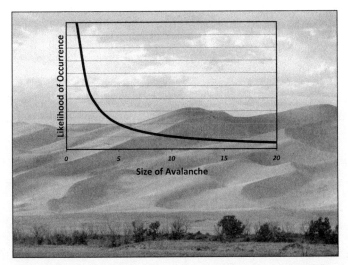

FIGURE 11.1: Random additions of sand eventually result in a self-organized critical system. Once the system achieves this critical state, additional sand can result in avalanches of any size, characterized by a power law distribution. *(Photograph by the author.)*

either subcritical (that is, adding a grain will just increase its height by one) or critical (that is, it's teetering in such a way that the addition of a single grain of sand will cause it to tumble onto a neighboring spot). Every grain of sand we add, every tumble that occurs, is continually pushing the system toward a critical state. At times, large swaths of the pile are poised so that the addition of a single grain of sand will cause an avalanche across the entire area. After that avalanche has devastated the pile, the system has relaxed enough so that new additions of sand either stay where they land or result in only small, localized avalanches that are quickly absorbed by subcritical neighbors. Overall, we observe long periods of relatively localized turmoil that increasingly drive the system toward widespread criticality, setting the stage for even a small event to trigger another widespread avalanche.

There is a relentless logic that drives such systems. Above, we assumed a simple physics with nicely behaved grains of sand that topple due to gravity when they pile too high. Even if we modify the physics by making the grains more irregular or by altering the force of gravity, similar behavior emerges. Under these new conditions, the system is still driven toward critical states. So whether we experiment with beach sand here on earth or dust on the moon, the self-organized criticality of the system remains a fundamental, emergent feature.

While the sand pile is dominated by simple physics, other systems may be driven by other mechanisms. For example, criticality in social systems might depend on features such as laws and regulations or financial risk. Laws and regulations may have little impact on social behavior at times. But as the circumstances of agents change, policies begin to bind, forcing agents into critical states where even small events can trigger large responses. Thus, we might see segments of society rise up against the government's taxation and fiscal policy, perhaps at first just forming local Tea Party–like movements, but on occasion triggering widespread social revolts. Or consider the banking and investment system, where various institutions try to maximize their returns by leveraging their assets and taking on risk. Over time, these systems can enter critical regimes where even small changes, perhaps the inability of one bank to repay a single loan, can result in a large avalanche as failure begets failure.

While in physical systems the drivers of criticality (such as gravity) are exogenous, in social systems they are often endogenous. Social elements such as tax rates and the amount of leverage banks are allowed to exercise are under the control of the government, typically through some political process. Political actors often have incentives to alter such policies in ways that might change the key determinants of criticality.

Consider a Classic-period Mayan city. The city proper is surrounded by farmers who must pay a tax to the government, either by turning over a share of their crops or by providing labor. In return, the farmers receive services from the city, such as protection, governance, and some insurance in case their crops fail. At low tax rates, the farmers are happy, because the amount they pay in taxes is more than compensated for by the services they receive. As the taxes are raised, the farmers become increasingly disgruntled with the trade-off they must make. At some point, things may become so bad that a farmer might rebel or move elsewhere.

Suppose our Mayan government, like most governments, prefers more revenue rather than less, perhaps because there is always a demand to build more elaborate temples. As the government raises tax rates, it starts to push the system closer to criticality. Every farmer is continually making a choice, weighing the benefits of staying at his current location against the tax he must pay. He'll consider his investments in improving the fields, his network of friends, his ancestral ties to the place, and so on. As the tax rate rises, the imbalance between the benefits of staying and the costs of leaving lessens, and the farmer is pushed closer to a critical state where even a small change—some bad weather or the loss of a cooperative neighbor, let alone a new government demand—could cause the farmer to up and leave.

If a farmer decides to leave, we see impacts that resemble those in our sand pile. On one hand, the farmer's field, now fallow and needing minimal investment to put it into production, might simply be taken over by someone else. Here, the departing farmer is like a grain of sand forming a subcritical hole in the pile. Alternatively, when the farmer leaves he might trigger his neighbors to leave as well. After all, the neighbors lost an important social connection who provided friendship and cooperation, and who by moving lessened the taboo against relocating the bones of one's ancestors. This latter situation is much like a grain of sand surrounded by other grains all in a critical state.

Unlike physical systems, social systems are likely to embody additional endogenous forces that could accelerate their criticality. For example, the immediate loss of production from the relocating Mayan farmer may force the government to increase its taxation on the remaining farmers. This will cause a further increase in system-wide criticality. Indeed, such endogenous drives toward increased criticality may be a natural outcome of social governance, as governments in pursuit of their goals tend to push citizens toward action. Once the system becomes critical, even trivial external events or policy changes can provoke system-wide reactions.

The idea of self-organized criticality may provide some needed insight into social phenomena involving rapid collapse and change. The rapid abandonment of Mayan cities in the Classic period could have been presaged by years of social policy that forced the system into a critical state. Once the society was in this state, the dynamics of the sand pile took over. Any social system is continually perturbed by seemingly insignificant events such as bouts of bad weather, missteps by the ruler, and so on. These perturbations usually have few noticeable consequences. Perhaps, on occasion, a farmer and maybe a neighbor or two decide to leave and set up operations elsewhere, but nothing much more than that. Yet such actions and reactions slowly flow through the system and inexorably drive it to a critical state. Once there, a seemingly minor affront to the system can trigger a large-scale avalanche.

On December 17, 2010, a Tunisian street vendor, Mohamed Bouazizi, set himself on fire to protest years of harassment by authorities. The event that triggered his protest was a municipal official publicly humiliating him by confiscating the scale he used to weigh his produce. Bouazizi tried to complain to the governor, but the governor refused to see him. This led him to the act that would eventually take his life.

Thus began the Arab Spring, where the confiscation of a vendor's scale in a rural Tunisian town started a wave of unrest that rippled outward from Tunisia into Algeria, Lebanon, Jordan, Mauritania, Sudan, Oman, Saudi Arabia, Egypt, Yemen, Iraq, Bahrain, Libya, Kuwait, Morocco, Western Sahara, Syria, and Israel's border towns. The outcome to date is a number of revolutions resulting in dramatic changes in governments, harsh crackdowns, and diplomatic maneuvering. The full impact of these events on the course of world history will likely be significant, but it's hard to fathom at this stage.

One can easily postulate forces, such as unhappy citizens or the dictates of an autocratic ruler, that could force a society into a critical state. Moreover, when one citizen is pushed so far that he decides to protest, this increases the likelihood that those nearby might take up the protest as well. Protests in various guises had been occurring in these countries for some time, but most of these efforts were quite localized. Nevertheless, they had been quietly driving the system to a more critical state. Once the system entered such a state, even an inconsequential act could trigger large-scale change, the consequences of which we are only starting to grasp. While such a hypothesis is speculative, one could test it by looking for the signatures of growing criticality in the various data feeds, such as Twitter, that may well have both captured and contributed to these events.

Self-organized criticality is an interesting form of complexity where small pieces of the system interact locally with one another, mediated by a very simple rule governing change. Over time, the system abstracts itself away from the particular local rule, and its global behavior is dominated by a characteristic pattern of avalanches at all scales. Most of these avalanches are small, but on rare occasions one encompasses the entire system. When global events occur, we want to invoke global causes. But the lesson from self-organized criticality is that there are forces underlying systems such that even small events, normally inconsequential, can have huge impacts.

At the slightest touch, our world can go from stones to sand.

TWELVE

From Neutrons to Life:
A Complex Trinity

As West and East

In all flatt Maps—and I am one—are one,

So death doth touch the Resurrection.

> —JOHN DONNE, "Hymn to God, My
> God, in My Sickness"

ON JULY 16, 1945, THE ATOMIC AGE BEGAN. AT JUST PAST 5:29 a.m. Mountain War Time, in the remote Jornada del Muerto basin (a desert in southern New Mexico aptly named by Spanish conquistadors for its "single day's journey of the dead man"), the Manhattan Project tested an implosion-initiated plutonium device that released the equivalent of around twenty kilotons of TNT. The test, conducted under the auspices of the project's scientific leader,

J. Robert Oppenheimer, was code-named Trinity, apparently derived from Oppenheimer's reading of two John Donne poems: "Hymn to God," which opens this chapter, and "Batter my heart, three person'd God." Three weeks later, an atomic bomb based on an untested but simpler design using uranium-235 was dropped on Hiroshima, Japan. Three days after that, a device based on the Trinity design was dropped on Nagasaki. Shortly thereafter, Japan surrendered, ending World War II.

Nuclear reactions, whether used for civilian power or for nuclear bombs, rely on interactions. In one type of reaction, energetic neutrons *potentially* collide with nearby nuclei, *perhaps* resulting in a fission event that releases some energy and even more energetic neutrons into the mix. Note the use of the words "potentially" and "perhaps." Chance plays an important role in such systems. If neutrons beget neutrons, there is the potential to transform mass into energy à la Einstein's famous $E = mc^2$. Depending on the speed of that transformation, we can get either warming leading to the carbon-free generation of civilian power or the destructive force of a nuclear blast. Given the potential energy inherent in an atom, it is not surprising that at the start of the war there was an intense interest in understanding the complex interactions that take place at this atomic scale. Such interactions embody the first branch of a complex trinity that leads us to a fundamental theorem about complex adaptive systems.

The second branch of our trinity begins with a decision to build a secret new device called the Electronic Numerical Integrator and Computer at the University of Pennsylvania's Moore School of Electrical Engineering. Called the ENIAC, it was the first programmable electronic computer, and it was a milestone development in the information age. The initial proposal to build ENIAC was made to the United States Army's Ballistics Research Laboratory at Aberdeen Proving Ground in Maryland by the physicist John Mauchly. Mauchly was inspired to find a better way to generate ballistic firing tables after encountering a group of women—known at the time as "computers"—literally cranking out such tables on desk calculators. Mauchly and an engineer, Presper Eckert, took charge of the ENIAC program, resulting in the eventual development of a thirty-ton electronic computer requiring 1,800 square feet of floor space, 17,500 vacuum tubes, and an astounding number of solder joints.

John von Neumann was a consultant to Aberdeen. Upon learning about ENIAC, he realized that it might be repurposed to help solve the "thermonuclear problem"—that is, a bomb predicated on nuclear fusion rather than fission—being pursued by some of von Neumann's Los Alamos colleagues led by physicist Edward Teller. In March 1945, von Neumann, Nick Metropolis, and Stan Frankel visited the Moore School and began to finalize plans for building a computational model of a thermonuclear reaction that would be run on ENIAC.

While the war ended before they had completed their work, by the spring of 1946 Metropolis and Frankel had discussed ENIAC with and presented their calculations and conclusions to a high-level group in Los Alamos, including von Neumann, Teller, Los Alamos director Norris Bradbury, Enrico Fermi, and Stan Ulam. Although the model was simple, the results were encouraging. In the annals of complex systems, this represents an important milestone in the use of electronic computation to understand complex interactions with serious, real-world implications.

Inspired by the results of Metropolis and Frankel, Ulam realized that a number of cumbersome but powerful statistical sampling techniques could be implemented using electronic computers. Ulam discussed this idea with von Neumann, who broached it with the leader of the Los Alamos Theoretical Division. The approach used computationally generated randomness to solve complex problems, and it marks the formal beginning of what has come to be known as the Monte Carlo method.* (The name was suggested by Metropolis, and it was "not unrelated to the fact that Stan [Ulam] had an uncle who would borrow money from relatives because he 'just had to go

---

* It appears that Enrico Fermi used Monte Carlo–like methods in the early thirties to solve problems in neutron diffusion. Apparently he enjoyed impressing his colleagues by offering quite accurate predictions of experimental outcomes based on clandestine mechanical calculations made during bouts of insomnia. There are also earlier examples of using randomness to perform important calculations, such as Buffon's needle from the eighteenth century being used to approximate the value of π.

to Monte Carlo.'") By the late 1940s the Monte Carlo method, promoted by various symposia, had become an accepted scientific tool.

Monte Carlo methods played a key role in a ground-breaking paper, "Equation of State Calculations by Fast Computing Machines," published in 1953. The paper had five authors: Metropolis, Arianna and Marshall Rosenbluth, and Augusta and Edward Teller. At the heart of the paper was "a general method, suitable for fast electronic computing machines, of calculating the properties of any substance which may be considered as composed of interacting individual molecules." This has become known as the Metropolis algorithm.

The paper focused on the question of how interacting particles will be distributed in space. Each particle interacts with the others, and we can calculate the overall energy for any given configuration. The challenge posed in the paper was to find the most likely configurations for this system.

One approach to solving this problem would be to model particles as coins and space as a tabletop. We could randomly toss coins on the table, calculate the energy of the resulting configuration as if the coins represented the position of atoms, and repeat. After numerous repetitions, we would begin to get a sense of the distribution of possible energy states for the system. Alas, the problem with this approach is that a lot of our effort will be spent on generating configurations that are not all that likely to appear. In physics, it is assumed that systems seek out lower-energy configurations, and a lot of our random tosses would result in high-energy configurations that we are not likely to observe.

Metropolis and colleagues provided an alternative solution for finding the most-likely configurations, and the technique embodied in this solution has profound implications both for modern statistical methods and, as we will see in the last branch of our trinity, for understanding complex adaptive systems.

The solution suggested by Metropolis and his coauthors seems remarkably simple on its face. First, begin with a randomly derived configuration of the coins and label this the status quo. Next, consider a new configuration of the coins generated by taking the status quo and randomly moving one of the coins by a small amount using a random process driven by a proposal distribution. We now calculate some measure of interest tied to the resulting configurations (in the above system, the amount of energy resulting from the interactions of atoms tied to the locations of the coins) for both the status quo and the new configuration. We then use an acceptance function to decide which of these two configurations will become the new status quo. If the candidate configuration is superior to the previous status quo, the candidate is accepted immediately as the new status quo. If the candidate configuration is inferior, then the likelihood that it replaces the previous status quo is proportional to their respective measures of interest. The more inferior the candidate is to the previous status quo, the less likely it will become the new status quo.

The algorithm proceeds by iterating the steps above using the current status quo. Amazingly, if we track the various configurations that this algorithm visits over time, we find that these configurations converge on the exact distribution underlying the measure of interest. That is, the system will spend relatively more time in those configurations that have the highest measure of interest. In the case of our particle system above, if we run the algorithm for a while and then randomly sample the status quo, we will tend to find the system in low-energy states.

Intuitively, the algorithm's behavior makes sense, as our acceptance criteria tend to direct the system into areas with higher measures of interest while avoiding areas with lower values. That being said, the fact that the system *perfectly* aligns with the distribution tied to the measure of interest is much more surprising, given that the algorithm never uses any global information about the underlying distribution.

However miraculous the algorithm's behavior may seem, it is possible to understand it mathematically. The first key piece is that the algorithm always uses the existing status quo as an anchor. The status quo embodies some important information, and thus the algorithm is not just randomly searching across possible configurations, which would result in a very different outcome. For example, we can forecast the weather by simply calculating the number of days that it rains throughout the year and then using this proportion as our prediction of rain on any given day. Alternatively, we could calculate the likelihood of rain on a particular day given that it rained on the previous day. As you might suspect, the latter approach gives us a very different set of probabilities and, in the case of weather, a much more accurate prediction, since the weather today is a good predictor of the weather tomorrow.

The idea that one event—say, yesterday's weather or the status quo configuration—might influence the probability of the next event goes back to the Russian mathematician Andrey Andreyevich Markov in the early 1900s, who developed a number of important results about these systems that are now known as Markov chains. The 1953 paper of Metropolis and colleagues used some of Markov's ideas to create a new class of algorithms called Markov chain Monte Carlo (MCMC) methods.

If the probabilities that govern the transitions from one state to another make it possible to go from every state to every other state (though not necessarily in one move), then the Markov chain is known as ergotic (or irreducible). If such transitions are always possible after n or more steps, the chain is called regular.* In the MCMC algorithm, it is easy to choose a proposal distribution for the next configuration that will guarantee that the resulting Markov chain will be regular. Typically, it also is useful to have a sym-

---

* All regular chains are ergotic, but not all ergotic chains are regular. For example, if you have a two-state system that alternates from one state to the other at each time step, it will be ergotic, because it is possible to go from any state to any other state, but not regular, since it visits a given state only on either even or odd time steps.

metric proposal distribution (as was done in the original Metropolis algorithm). This requires that the probability of proposing x given y is identical to the probability of proposing y given x.

The fact that MCMC algorithms are driven by a regular Markov chain is useful, as these chains become increasingly well behaved as the system runs forward. Such systems collapse into a very well-behaved regime where the probability of finding it in any particular state is fixed and independent of where the system started. That is, if we run our MCMC algorithm long enough, the system will start to visit the various states in a predictable way. In the case of weather, regardless of the weather today, if we wait long enough, the chance of rain on a randomly chosen day in the future will be fixed. These systems do, however, require a certain amount of burn-in time. While we can start the system anywhere, it takes the chain a certain amount of time to forget where it began and to find its more fundamental behavior. Exactly how long that takes has important implications for adaptation.

The second key aspect of the MCMC's remarkable behavior involves the acceptance criteria. The choice of the acceptance criteria is made with careful forethought, as it guarantees that the algorithm converges to the underlying probability distribution implied by our measure of interest. The odd thing about this convergence is that we typically cannot calculate this probability distribution directly, as it requires information about all possible configurations of

the system, and the number of such configurations is usually so large that performing this calculation is impossible.
Fortunately, the algorithm doesn't need to directly perform
such a calculation. A sufficient condition that guarantees
the needed convergence is a property known as detailed
balance. Detailed balance requires that when the system
converges, the resulting transitions are reversible in the
sense that the equilibrium probability of moving from one
configuration to another is the same in either direction. If
detailed balance holds, then the system converges to the
underlying probability distribution driving the measure of
interest.

Arthur C. Clarke noted that "any sufficiently advanced
technology is indistinguishable from magic," and perhaps
MCMC is one such technology. Through a very simple
set of manipulations—randomly creating a candidate configuration contingent on the status quo and possibly replacing the status quo with that candidate configuration
based on a roll of the dice tied to the relative measures of
interest of the two configurations—we create a system that
is driven by a much deeper force, one that was previously
inaccessible to us given the impossibility of making the required calculations. MCMC methods have radically altered
the scientific landscape. In particular, such algorithms are
needed for the widespread application of Bayesian statistics, a now ubiquitous approach that is ushering in a new
analytic age characterized by everything from targeted web
ads to driverless cars. At the heart of Bayesian methods is

the need to calculate some key probability distributions. Such distributions are often computationally impossible to calculate directly, but we can invoke the magic of MCMC.

This brings us to the third branch of our complex trinity, namely, the implications of the first two branches for understanding complex adaptive systems. The impetus for the original development of MCMC algorithms was to find a simple method that could reveal a previously hidden but vitally important distribution. In the case of Metropolis and colleagues, this was the distribution of the likely energy states for a system of interacting particles. To accomplish this goal, the creators of the algorithm developed a set of simple, iterative steps that generated the needed candidates. As if by magic, such simple steps are sufficient to uncover the desired distribution. The implications of MCMC algorithms are even more profound for complex adaptive systems. The mechanisms that drive adaptive systems, such as evolution, have direct analogs to the key elements of the MCMC algorithm. In short, a complex adaptive system behaves as if it were implementing a MCMC algorithm.

Consider a pond covered with lily pads, on one of which sits a frog. We assume that each lily pad, depending on its location, has some inherent value for the frog—say, the number of flying insects that the frog can snatch in a given time (assuming that the supply of such insects is constantly renewed). Suppose the frog behaves as follows: Every minute it randomly picks a neighboring lily pad, and if the number of insects at that new pad is greater than the number at the current pad, it jumps to the neighboring pad. Otherwise, it jumps to the new pad with a probability proportionate to the relative number of insects at the two locations. Thus the chance of moving will be higher the closer the number of insects at the neighboring pad is to the number at the current one.

As might be apparent, the frog is in the midst of a MCMC algorithm. The lily pads represent various states of the system, and the location of the frog marks the status quo configuration. When the frog considers a randomly chosen neighboring lily pad, it is drawing from a proposal distribution. The frog's decision to jump to the new pad (that is, to create a new status quo) is implemented via the acceptance criteria used in the Metropolis algorithm, always accepting the new pad (state) if it is better, and if it is worse, accepting it with a probability tied to the relative quality.

Given that the frog is implementing a MCMC algorithm, we can easily characterize its long-run behavior. After a certain amount of initial hopping (aka burn-in time), if we track the frog's location, we will find that the amount of time it spends on any particular lily pad is given by the number of insects found at that pad divided by the total number of insects across all of the pads. These fractions, taken together, give us a probability distribution describing the likelihood of the frog being on a particular lily pad sometime in the future. For example, if we have three lily pads, the first one with fifty insects, the second one with thirty, and the third one with twenty, then over time the

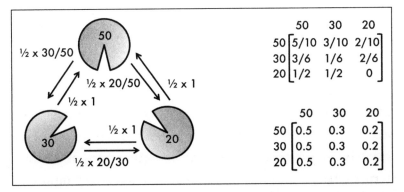

FIGURE 12.1: Three lily pads with different numbers of flying insects given by the numeric label in each pad. We assume that at each time step the frog has an equal chance of picking one of the two neighboring lily pads and moving to it based on the Metropolis acceptance criteria. The diagram on the left illustrates the overall design, with the labeled arcs providing the probability of choosing that neighbor (½ in all cases) × the acceptance probability. The top matrix on the right gives the transition probabilities for the resulting Markov chain Monte Carlo process, that is, the probability of moving to the pad designated by the column, given that you are currently on the pad designated by the row. After many time steps, the system converges to the bottom transition matrix, with the frog residing on the 50-, 30-, and 20-insect lily pads 50 percent, 30 percent, and 20 percent of the time, respectively. This bottom matrix is the result of repeatedly multiplying the top matrix by itself.

frog will be on the first pad 50 percent of the time, on the second one 30 percent of the time, and on the third one 20 percent of the time (see Figure 12.1).

What's true for frogs and ponds can also be true for other adaptive systems. Configurations of an agent in an adaptive system represent various states of the system. An agent pursues goals by reconfiguring itself in constrained ways and moving toward new configurations that result in better outcomes. So if we are willing to make some key simplifications of reality, we can generate a useful model of an adaptive agent in a complex system and use MCMC ideas to derive a fundamental theorem about this system's behavior.

To begin this process, we need to think about how an agent represents various states. In biology, for example, an organism's genotype represents a possible state of the system of all genotypes. In economics, states can represent, say, the standard operating procedures of a firm, the design of a product, the consumption bundle of a consumer, or some rule of behavior. In a city, states could represent the network of roads or the locations of activities.

Given the state space of an adaptive system, the MCMC approach requires a notion of identifying new possible states using a proposal distribution. For a well-behaved MCMC, the requirements for the form of the proposal distribution are relatively light—not much more than some reasonable ability to move, eventually, from one state to another. (For convenience, we might want to impose some more restrictions, such as requiring that this distribution be symmetric.) In adaptive systems, the notion that existing structures can be randomly altered by a mutation-like operator tends to be an easy assumption to embrace and one that meets the above requirements. Indeed, mutation is central to biological systems, and it also approximates the behavior observed in a variety of other systems. For example, new products are often slight variants of old ones, new technological or scientific ideas rest on the shoulders of giants, consumers make slight alterations in the goods that they buy, and so on.

The algorithm needs a measure of fitness for any given state of the system. For our frog, insects serve as this measure, as presumably the more insects the frog consumes, the happier it is. Similar measures of fitness arise in other adaptive systems. For example, in biology there is a notion of fitness that is tied not only to the food supply but also to the overall ability of an organism to survive and reproduce. In economics, we often think of agents pursuing profit (in the case of firms) or happiness (in the case of consumers). Thus, as long as agents in an adaptive system are pursuing a goal, we can use the measure of this goal as a means to drive our model.

Finally, our model requires that the adaptive system adopt the proposed variant based on a MCMC-compatible acceptance criteria. The criteria developed by Metropolis and his coworkers always accepted any variant that had higher fitness than the status quo, and when the fitness of that variant was less than that of the status quo, the criteria adopted it probabilistically, with the likelihood diminishing as the difference in fitness between the two options increased. Such a rule seems to be a reasonable approximation for many adaptive systems.

There may be other acceptance criteria that will also result in detailed balance and thereby provide the needed convergence. For example, if the proposal distribution is symmetric, an acceptance criterion that adopts the variant with a probability given by the variant's fitness divided by the fitness of both the variant and the status quo also results in detailed balance. Under such a rule, if the value of the variant is equal to that of the status quo, there is a 50 percent chance that it is adopted, and as the variant's value rises (falls) the adoption probability increases (decreases) away from 50 percent. (By the way, all is not lost if the system has acceptance criteria without detailed balance, as it will still converge to a unique distribution, but that distribution will differ from the one specified below, though perhaps only slightly.)

Suppose we have an adaptive system where an agent represents a possible state of the system. At each time step, a reasonable variant of this agent is tested against the environment, and that variant replaces the existing agent based on detailed-balance-compatible acceptance criteria tied to relative fitness. This leads to a fundamental theorem of complex adaptive systems: in the above adaptive system, after sufficient burn-in time, the distribution of the agent (states) in the system is given by the normalized fitness distribution.

The proof of this theorem is simply noting that the above system is implementing a MCMC algorithm, and that such algorithms converge to the (implicitly) normalized distribution of the measure of interest—here, the fitness used in the acceptance criteria.

This theorem implies that, in general, such adaptive systems converge to a distribution of states governed by the normalized fitness. Thus, adaptive agents are not perfectly able to seek out, and remain on, the best solutions to the problems they face. Rather, they tend to concentrate on the better solutions (given sufficient time), though on rare occasions they will find themselves in suboptimal circumstances. If we throw our frog into the pond and give it a chance to adapt for a while, when we return the frog is most likely to be on the lily pad with the most insects, but there is always a chance that we will find it on any of the other lily pads—a lower chance for those lily pads with fewer insects, but a chance nonetheless.

One implication of the theorem is that while adaptive agents tend to do well, they are not perfect. Such a statement is both comforting and disconcerting. While it is nice to know that, given sufficient time, adaptive systems tend to concentrate on the fitter outcomes in the world, on rarer occasions they will end up in the bad outcomes. While the acceptance criteria tend to bias movement toward better parts of the space, there is always a chance that the system will move from a high-fitness outcome to a low-fitness one.

The obvious question here is whether the system would be better off by avoiding such fitness-decreasing moves. By preventing such moves, we could ensure that the algorithm always walks toward areas of higher fitness but, as we saw in Chapter 5, such an algorithm can easily get caught at a local maximum—a point where all roads lead downhill even though much higher terrain exists in the distance. Thus, the need to accept configurations of lower fitness is a necessary evil that prevents the system from getting stuck on local maxima.

One could modify the algorithm by, say, introducing a temperature, as is done in simulated annealing. Early on in the process, the temperature is kept high, allowing the algorithm to proceed normally. As time passes, we cool the search, lessening the chance of accepting fitness-reducing moves. Given enough time and a carefully controlled annealing schedule, the system will tend to lock into areas of higher fitness. But the design of such annealing schedules is tricky, and once the system is cooled, it would be unable to adapt to a change in the underlying fitness landscape.

Note that the theorem requires sufficient burn-in time. Recall that Markov chains link the probability of the next state to the current state. In this sense they have some memory, since where you are now influences where you can go in the short run. Under the right conditions (which hold in our theorem), these initial influences dissolve away over time and the subsequent links in the chain are driven by the more fundamental forces characterizing the Markov process. Burn-in is the amount of time it takes for the system to forget its initial conditions and fall into its fundamental distribution. The needed burn-in time depends on a number of factors. As we increase the size of the underlying space, the burn-in time increases, since it takes longer to explore the larger space. Moreover, the burn-in time can be influenced by our proposal distribution. If the proposal distribution produces variants that are very close to the status quo, then the Markov chain will be slow to form, given the plodding search. If instead the variants are far from the status quo, then rejections are quite likely, and again the chain will slow down. Finally, the shape of the space itself can influence burn-in time. For example, if there are large areas of low fitness, the chain can get caught in these desolate flatlands for long periods of time before it happens upon the fitter parts of the space.

Unfortunately, other than the intuitive arguments above, our ability to succinctly characterize the burn-in process at a theoretical level is quite limited. Nonetheless, burn-in has some interesting implications for adaptive systems. While our theorem guarantees that the adaptive system will *eventually* fall into the normalized fitness distribution, the speed at which this happens depends on how quickly it can traverse the burn-in period. Systems with larger state spaces, more anomalous fitness landscapes, particularly bad starting conditions, or proposal distributions generating variants that are either too near or too far will tend to hamper the ability of adaptation to quickly converge on the fitter variants governed by the normalized fitness distribution. In this sense, longer burn-in times make adaptation more difficult.

The above theorem, like all theorems, was predicated on a set of simplifications. It assumes that adaptive systems work by marching from one structure to the next, with new structures being generated via a proposal distribution and being accepted based on acceptance criteria that are tied to the relative fitness of the proposed variant. This is a somewhat static model, as the fitness distribution is unchanging in the sense that the identical structure gets the same measure of fitness in perpetuity. In more realistic systems, one might want to include an endogenous notion of fitness, whereby the fitness of a given structure depends on, say, what other structures are in the world. This may be possible by extending the notion of structure in the model. Instead of thinking of it as defining a single agent, it could define an entire population of agents, but such an elaboration is non-trivial, since the fitness function is typically defined at the level of the individual, not at the group level. In such an extended system, it is as if we are running multiple MCMC algorithms, with each agent adapting to the (transient) world created by the other structures.

The other two key elements driving our theorem are the proposal distribution and the acceptance criteria. MCMC algorithms tend to be fairly robust if the choice of the proposal distribution is reasonable, though, as noted, this choice may influence burn-in time. Unfortunately, clean theoretical results about the relationship between proposal distributions and burn-in time are difficult to derive, apart from the notion that tuning the distance of the search can influence burn-in time, with a Goldilocks-like region of jumps that are neither too large nor too small leading to the fastest burn-in time.

The acceptance criteria are another interesting element of the algorithm. The original acceptance criteria used in the Metropolis algorithm were driven by necessity of design, and while they provide a reasonable analog to a lot of adaptive processes, other criteria may be of interest. There are alternative acceptance functions, for example, some that use a more direct measure of relative fitness that can also result in detailed balance, thereby allowing the system to converge as per the theorem. Having a better characterization of the class of acceptance criteria that results in detailed balance would be useful. Also, even when detailed balance doesn't hold, the system still converges to a unique state distribution, but that distribution will not be given by the normalized fitness. In these cases it is still likely that the alternative acceptance criteria will lead to system behavior that approximates the results above or, alternatively, have interesting implications of their own.

WE BEGAN THIS CHAPTER WITH THE EXIGENCIES OF WAR, which resulted in the need to understand interacting atomic systems and to develop novel tools (such as programmable computers) and methods (such as Monte Carlo) that could be used to provide essential insights into such systems. Taking place at the dawn of both the atomic and information ages, it is a story of great genius and, at times, even playfulness. Our complex trinity was completed by repurposing the algorithms of war to gain some fundamental insights into the behavior of an adaptive agent in a complex system. We find that such agents are unknowingly implementing an algorithm that locks them into a cosmic dance of fitness. As Metropolis noted, "What a pity that war seems necessary to launch such revolutionary scientific endeavors."

Epilogue:
The Learn'd Astronomer

When I heard the learn'd astronomer,

When the proofs, the figures, were ranged in columns before me,

When I was shown the charts and the diagrams, to add, divide, and measure them,

When I sitting heard the astronomer where he lectured with much applause in the lecture-room,

How soon unaccountable I became tired and sick,

Till rising and gliding out I wander'd off by myself,

In the mystical moist night-air, and from time to time,

Look'd up in perfect silence at the stars.

—WALT WHITMAN, *Leaves of Grass*

W E HAVE ALL HEARD THE LEARN'D ASTRONOMER IN ONE FORM
or another. That is, we have encountered a carefully

laid out analysis that, while perhaps worthy of applause, seems terribly disconnected from the stars we wish to gaze upon and know.

A similar discomfort is pervasive in modern academics. We have carefully worked out analyses of, say, the chemical interactions in the brain or the optimal bidding strategy in a simplified auction, yet the connection from these studies to the phenomena that truly awe us, such as the ability of brains to think or markets to organize trades, seems rather specious.

While we might want to blame the scientists for caring about the wrong things, it's not that easy. The reductionist approach to the world—breaking down complicated things to their constituent parts and then carefully dissecting these parts until we know them—has provided a useful Archimedean purchase from which to lever our complex world into the light of understanding. Unfortunately, there is only so far we can move the world using such tools.

As we have seen throughout this book, knowing the parts is not equivalent to knowing the whole. Reduction does not tell us about construction. This is the fundamental insight of the study of complex systems. Even if we could fully understand how an individual worker bee's, market trader's, or neuron's behavior is determined by its environment, we would have little idea about how a hive, market, or brain works. To truly understand hives, markets, and brains, we need to understand how the *interactions* of honeybees, traders, and neurons result in system-wide, aggregate behavior. Take a single worker bee and fully analyze how she responds to the chemical, visual, and audio inputs she receives, and we gain some new knowledge about how a simple organism responds to its world. Take a hive of such honeybees and allow them to interact, and we begin to see new behaviors emerge that, while obviously tied to the actions of each individual honeybee, are at the same time wholly disconnected from these actions and not easily predicted from only our observations of the individual worker. Like other biological organisms that we know (including us), this newly emerged entity has the ability to regulate its own temperature, gather needed nutrients, store and use energy, protect itself from outsiders, dispose of waste, attack internal and external threats, and even reproduce.

The ultimate hope in the science of complex systems is that honeybee hives, financial markets, and brains are deeply connected—or, for that matter, not all that different from other biological organisms, cities, companies, political systems, computer networks, and on and on. A honeybee swarm may just be a more easily observed instance of a brain. If so, the general processes of positive feedback for better choices and quorum-based triggers to finalize the decision may drive not only a swarm's choices but also our own.

Over the last two decades, various strands of complex-systems thinking have slowly come together in an emerging tapestry of understanding that considers not just one particular thing but the whole. The warp of this tapestry, stretched tight on the loom of science, is composed of the key ideas and tools that have become integral to the study of complex systems. The weft, weaving through the warp and binding it together, is beginning to fill in a slowly emerging pattern. There are many weavers working on this tapestry, each attempting to keep coherence and beauty in her own part of the work. Recently we have been starting to see these various pieces begin to meld into one another, marked only by faint lazy lines like those seen in Navajo rugs. The various strands of complex-systems ideas and examples explored here are starting to create a beautiful, and quite useful, tapestry.

WE BEGAN OUR EXPLORATION INTO COMPLEX SYSTEMS BY looking at how simple, local actions, once connected, can result in new global patterns. These types of systems abound in our world, and we now know that from simple beginnings we can get marvelous ends. Small pieces of colored glass, once connected, result in a stained-glass window that creates an image in our mind's eye that inspires faith and spirituality. Even now, as you read these words, pixels are becoming letters, letters are becoming words, words are developing meaning, and meaning is rolling into thought.

To understand how simple parts result in global patterns, we used the mathematics of cellular automata. These abstract creations, linked with the utility of a computer to help visualize their implications, show how simple, local, decentralized processes can result in global patterns.

More than two hundred years ago, Adam Smith invoked the "invisible hand"—not far removed from "then a miracle happens"—to explain how the individual actions of traders, each out for her own gain, result in an outcome that was no part of anyone's intention. The complex-systems perspective begins to make visible Smith's guiding hand as we investigate the trades that arise in a bustling bazaar. Markets, when conditions are right, leverage the power of simple beginnings to lead to emerging global patterns (prices) that allocate resources and economic production to best use.

When systems are simple, we can easily trace their behavior from step to step, and in so doing make accurate predictions about the overall system's behavior. When systems are complex, such tracing becomes far more difficult, as each new trace alters previous ones, making it extremely tricky and at times impossible to predict what will happen—as when, on May 6, 2010, a computer in Shawnee Mission, Kansas, compounded a seemingly insignificant error in its trading program and created an unanticipated feedback loop that wreaked havoc on the financial shores.

Anytime we interconnect systems, we build in feedback loops. Some types of feedback result in stabilizing forces, calming the system as a whole. Alas, other types of feedback destabilize systems, and even with careful thinking and design, it is easy to build systems with unintentional—and unfortunate—feedback loops. The worldwide financial collapse of 2008, the reverberations of which are still playing out today, was the result of a system in which each of the parts seemed rational—or at least responded in a reasonable way to local incentives. Unfortunately, what was true for the parts was not true for the whole, and the interconnections among the parts resulted in a series of feedback loops able to take down a worldwide economy. We often hear that some disastrous event was the result of a "perfect storm" of events. However, in a world of increasing complexity, we are simply perfecting our ability to create such storms.

When complexity abounds, diversity matters. Systems composed of homogeneous agents behave quite differently than those composed of heterogeneous ones. Homogeneous systems, with all of the agents taking the same actions based on the same cues, can have much more dramatic responses to new events than heterogeneous ones. Thus, if we want to predict better how a system will behave, we need to account explicitly for its heterogeneity rather than rely on theoretical expediencies such as representative (and hence homogeneous) agents.

The value of heterogeneity depends on the system we are considering. Heterogeneity is useful when a graduated response is needed, and so in systems like honeybees attempting to control their hive's temperature or traders wanting to stabilize prices, having more heterogeneity is better. However, there are other systems where such heterogeneity is detrimental. A government wanting to control a social movement or a population of bacteria wanting to release a toxin might benefit from less heterogeneity.

When complexity abounds, discovering solutions to problems is often difficult. The various interdependencies and feedbacks that are innate in a complex system make searching such systems for new solutions extremely hard. In less complex parts of our world, finding good answers is like climbing Mount Fuji, where if you just keep going uphill, you will reach the top. In such a world, errors—taking a step in the wrong direction—only hamper progress. Complex systems involve a very different kind of search, where the mountain range is not only rugged but also fog-bound, and perhaps even undulating with every step we take. Such complex ranges are common when we have interconnected and interdependent pieces making up the whole, such as we might see in consumer goods, technologies, manufacturing processes, and drug cocktails. In such a world, even if we hike up the fog-bound hill without error, we might miss the highest point on the landscape. Indeed, we might find ourselves victoriously standing on top of a molehill, falsely thinking that we have made it to the top of the mountain. To avoid making mountains out of molehills, we need new ways to search in complex worlds. In particular, introducing errors into our search process—that is, occasionally taking random steps downhill—may allow us to escape the trap of the molehill and head to the mountaintop.

When complexity abounds, decision making becomes ubiquitous. As thinking beings, we find it all too easy to believe that decision making requires intelligence, and that intelligence requires a brain. Yet in a world of complex connections and interactions, simple parts can result in intelligent decisions.

From the white blood cells in our immune systems to the multitude of bacteria inhabiting our bodies, trillions upon trillions of intelligent decisions are being made every second, without a neuron to be had. We live in a sea of computation and decision making. Nature has, by linking together simple chemical and physical processes, created biological computers capable of taking input from the world, remembering it, and acting upon it in useful ways.

Such biological decision-making systems, shaped by evolution and operating without a neuron-driven brain, are capable of making intelligent choices. Remarkably, even lowly bacteria are taking actions driven by well-defined sets of preferences. More surprising is that, like their big-brained human brethren, these simple systems also fall prey to various biases in decision making that result in suboptimal performance. Neurons are useful in that they can quickly transmit signals across (biologically) large distances, but of course many organisms are not all that big. Once we give up the need for neurons, the decision-making processes used by a bacterium and a human may not be all that different. We may well exist in a world where thinking is everywhere.

If individual decision making and intelligence do not require the existence of a brain, then it is not too far a leap to think about how such intelligence could emerge in collections of agents. For example, a honeybee worker has a limited set of behaviors tied to cleaning the hive, nursing brood, forming comb, collecting nectar and pollen, and so on. However, place thousands of such workers into a colony, and we get an entirely new set of useful, colony-level behaviors. Out of the interactions of thousands of honeybees emerges a superorganism capable of a survival-tested behavioral repertoire. And what is true for a collection of honeybees also holds for collections of other types of agents. Each year, tens of thousands of recorded songs are released, and society attempts to identify the better ones via top-ten lists and the like. It is not hard to take the insights we gained from swarming honeybees and apply them to the *Billboard* Hot 100 list, as the likelihood of a particular song making it onto the list is indirectly tied to its quality and to how often it gets heard. Political primaries and public debates are subject to similar mechanisms.

One of the most intriguing links to group decision making is to our own consciousness. Neurons in the brain and honeybees in a colony may not be all that different, and if so, we have a new way to think about thinking. Following such a path suggests that the mind of the hive that emerges in honeybees may explain the hive of our own mind.

When complexity abounds, networks of connections matter. These networks, determining the interaction possibilities of agents, result in the emergence of global patterns. Sometimes these patterns are useful, as in the case of swarming honeybees finding a good location for the next hive. Other times, networks result in undesirable outcomes. For example, even if we begin with people who have only a slight preference to live among similar types of people, we can easily end up with a highly segregated society. What is true for a neighborhood is also true for political choices, religious beliefs, crime, and other social norms. We live in a world where even good intentions can easily get overwhelmed as complexity plays out over the network, forcing us to an outcome that no one intended or desired.

When complexity abounds, scaling laws may prevail. We are only slowly uncovering the potential laws that may govern the various complex systems we inhabit. Knowing the population of the largest city in a country can tell us the population of the second-largest city. Knowing the heart rate of a mouse can tell us the life span of an elephant. Knowing the number of wars in which a thousand people died can tell us the number of wars in which a million will perish.

The existence of scaling laws is not only empirically convenient but also theoretically suggestive of a deeper unification among systems. The same approach that allows us to derive scaling laws for biological systems may also work for social and artificial systems as well. Cities are organisms that require energy to be transported, stored, and used, and thus cities might exhibit scaling laws akin to biological systems. The key demographic trends over the last century have been continued population growth and increased urbanization. We now exist in a world of more than 7 billion people, with more than half of them living in urban areas. Knowing the power laws of cities will give us key insights into our future prospects on this planet.

When complexity abounds, cooperation can emerge. The ability to cooperate is a key element in the success of our species. In most social worlds, competition makes you slightly better off, while cooperation makes you remarkably better off. Unfortunately, individual incentives tend to favor competition over cooperation.

Notwithstanding a world red in tooth and claw predicted by individual incentives, there are enough examples of cooperation emerging in complex systems to provide a ray of hope. Rice farming on the island of Bali occurs under conditions that seemingly favor the suboptimal, competitive outcome. Yet as the human and natural ecosystems became more tightly coupled, farmers began to cooperate and coordinate their farming activities, resulting in more food for all.

Cooperation appears to emerge in other systems as well, despite what appear to be compelling reasons for competition to prevail. Using abstract models of evolving computer programs, we can explore the origins of cooperation. In such worlds, cooperation emerges when evolving strategies repurpose their early plays of the game and learn to communicate with one another and signal a willingness to cooperate. When such signals are sent and acted upon, something like a secret handshake spontaneously arises as a way to distinguish self from other, and cooperation can thrive.

When complexity abounds, self-organized criticality can arise. Complex systems often organize themselves into characteristic configurations that embody unintentional order. This order implies a system that is near the edge of action. Self-organizing critical systems result in a world where activity can occur across all scales. The majority of events tend to result in small, localized avalanches, though rarely a small event can result in an avalanche that encompasses the entire pile.

Certain types of social systems may self-organize into critical states as well. In these systems, small actions that normally have little consequence, such as the protest of a desperate Tunisian street vendor in a remote town, can at times trigger large results, such as the subsequent wave of government overthrows across the Middle East starting in late 2010, known as the Arab Spring.

Complexity abounds even in our attempts to understand complexity. The desire to harness the atom for war led to a remarkable web of interacting people, ideas, and technologies. From this web emerged a Promethean bargain, creating not only the most destructive weapons humankind has ever known but also a core set of ideas and tools that have set the stage for the modern science of complex systems. By cleverly harnessing the computational substate that arose during this period, we have been able to rapidly advance our understanding of interacting systems over the past few decades.

Indeed, algorithms originally designed to understand the behavior of atoms, and currently used to drive our emerging age of analytics, provide a new view of complex life. Adaptive agents are part of a cosmic algorithmic dance, subject to deep forces that determine their fate.

ULTIMATELY, COMPLEXITY ABOUNDS IN THE CHALLENGES that we face as a society. Take any of the major issues confronting humanity—climate change, financial collapse, ecosystem survival, terrorism, disease epidemics, social revolution—and you will see that they have a grounding in complex systems. Ideas from complex systems are starting to reshape the way we think about and act on our world. Policies that follow the reductionist model—for example, considering only the securities held by a single bank while ignoring the interconnections and codependencies that bind banks together—are doomed to fail. It is only by embracing the broader complex-systems perspective that policy making can keep up with our increasingly complex world.

The various strands of ideas, theories, and observations explored here are forming an important new tapestry that will give us fresh perspectives on our world and provide novel ways to advance our goals. The emerging tapestry of our understanding of complex systems is itself subject to the laws of complex systems. Thus, it is taking on a global beauty, coherence, and utility that were no part of any individual weaver's intention or ability. The various proofs, observations, and conclusions that form each thread are beginning to fade into a deeper understanding as we look, in perfect silence, at the complexity that abounds.

Index

allometry, 160
Anderson, Phil, 2
ants
  army, 133–134
  circular mill, 134
  colony finding, 127
  resource decisions, 133–134
  tandem running, 127
Arab Spring, 18, 201–202, 238
Arrow, Ken, 30
automata, 28, 183–193
  communication, 188–193

Bak, Per, 196
Bali
  rice farming, 16–17,
    169–178, 237
  time of *poso*, 173
  water temples, 171–173, 175

Barrett, Brittany, 93
basins of attraction, 83–84
bazaar economics, 39–44
Beekman, Madeleine, 136
Bettencourt, Luis, 167
Borges, Jorge Luis, 2, 195
Bouazizi, Mohamed, 201
Boulding, Kenneth, 43
Bradbury, Norris, 208
Brown, Jim, 161
burn-in, *see* Markov chain

cellular automata, 23–30, 230
  Rule 22, 29
  Rule 30, 24–29
chemotaxis, 101–105, 109
Cities and the Wealth of Nations,
    70–71
Clark, Arthur C., 214

241

competitive equilibrium, 35–45
computing machines, *see*
    automata
cone snail, 25–28
    Conus omaria, 26
cooperation, 16–17, 169–198,
    237–238

Darwin, Charles, 11, 21, 185
Debreu, Gerard, 30
decision making, 12–13,
    106–108, 233–235
    Billboard music charts, 118,
    127, 235
    decentralized, 12–13, 113–139
    errors, 79–80, 87–89,
    131–132
    group, 22, 113–139
    macaque monkeys, 131–133
    molecular, 18, 100–111
    political primaries, 235
    visual, 131
detailed balance, *see* Markov
    process
Donne, John, 205–206
Double Auction Tournament,
    55–56
drug cocktails, 86–87, 90–98
    cancer chemotherapy, 90,
    92–97

E-mini, 50–51, 53, 57–59
Eckert, Presper, 207

Einstein, Albert, 206
emergence, 68, 138–139
ENIAC, 207–208
Enquist, Brian, 161
errors, *see* noise
expectations in financial
    markets, 67–68
externalities, 170–172

feedback, 7–9, 16, 54, 58–59,
    61, 231–233
Fermi, Enrico, 208
flash crash, 8–9, 47–54, 64–67,
    172
Frankel, Stan, 207–208

Gell-Mann, Murray, 4, 164
general equilibrium, 31
genetic algorithm, 184–185
Goldschmidt, Richard, 185
Golman, Russel, 121
Gordon, Deborah, 133
group intelligence, 113–135

Hagmann, David, 121
Hanneman, Robert, 154
Hayek, Friedrich, 37–38
heterogeneity, 7, 10, 12, 70–73,
    80, 232–233
high frequency trading, 54–59,
    61–63
Hobbes, Thomas, 170–171
homogeneity, *see* heterogeneity

honeybees, 13–14, 229, 232
  hive finding, 113–126
  quorum, 119, 122–124
  superorganism, 113–115, 235
  swarms, 116–123, 129
  temperature regulation, 71–74
Hoyle, Fred, 189

intelligent design, 26
interactions, 6–8, 10–11, 27–28,
    228–229, 232
invisible hand, 5, 30, 37, 68, 231

Jacobs, Jane, 70–71
Joyce, James, 79

Kleiber, Max, 160

Lakeland, 141–151, 155
landscape, 1–2, 80–86, 87–89
  ruggedness, 83–87, 89, 94, 98
Latty, Tanya, 106
Lobo, José, 167

majority rule, 131, 143–145,
    147, 166, 238
maps, 1–2, 4, 19, 32, 34, 80, 82
markets, 8–10, 30–39, 41–45,
    47–62, 228, 231–232
  auctions, 55–56, 136–138,
    228
  self-organization, 140
  trader heterogeneity, 74–75

Markov, Andrey Andreyevich,
    212
Markov chain, 212–213,
    222–223
  burn-in, 213, 216, 220,
    222–224
Markov chain Monte Carlo
    (MCMC)
  method, 212–220, 224–225
Markov process
  detailed balance, 214, 225
  ergotic, 212
  irreducible, 212
  regular, 212–213
Mauchly, John, 207
MCMC algorithms, *see* Markov
    chain Monte Carlo
    (MCMC) method
Metropolis, Nick, 207–210,
    215, 219
molecular intelligence,
    100–111
Monte Carlo method, 208–209
more is different, 22

networks, 14, 16, 141–156,
    235–236
  crime, 167–168, 178, 236
  education, 148
  message passing, 149–152
  religion, 148
  six degrees of separation,
    151–152

networks *(continued)*
  small world, 150
  social policy, 201
neurons, 100, 129, 205–206,
    208, 234
neutral drift, 190–191
neutrophil granulocyte, 99
Newsome, William, 131
noise, 5, 7, 80, 87–88

Ockham, Father William of, 9
"On Exactitude in Science," 2
Oppenheimer, J. Robert,
    205–206
optimization, *see* search
order book, 51–52, 60

Pfeffer, Wilhelm, 105–106
policy, xviii–xix, 173, 198, 201
  heterogenity, 73
power law, 158–168, 196–197
prisoner's dilemma, 178–181,
    183

randomness, *see* noise
reductionism, 2–5, 9, 12, 22,
    62, 228
representative agent, 9–10,
    69–70
Richardson, Lewis Fry,
    164–165
risk aversion, 125–126
Robert's Rules of Order, 138

Rosenbluth, Arianna, 209
Rosenbluth, Marshall, 209

scaling, 15–16, 158–167, 236
  city size, 166–168
  firm size, 166
  life span, 157, 160, 163
  metabolism, 160–163
  population growth, 166,
    236–237
  urbanization, 168
  war, 164–165
  words in a text, 165–166
  Zipf's law, 165–166
Schelling, Thomas, 152
search, 80–95, 233
  hill climbing, 80–86, 88–89
  simulated annealing, 88–89,
    222
segregation, 152–155
self-organized criticality, 17,
    195–203, 238
  collapse, 195–198
  Maya, 195, 199–201
  revolts, 198
  social systems, 198–202
seven deadly sins, 64
Six Sigma, 79–80, 87, 98
slime mold, 106–107, 109, 111
Smith, Adam, 5, 30–31, 68, 231
Snell, Otto, 106
social movements
  heterogeneity, 75–77

Statistics of Deadly Quarrels, 164
Strumsky, Deborah, 167
supply and demand, 31–38

Taymiyyah, Ibn, 31
Teller, Augusta, 209
Teller, Edward, 207–209
theorem of adaptive systems,
    18–19, 215, 217–221, 223
Tolkien, J. R. R., 69
trading algorithms, 52–54,
    57–58, 77
Trinity test, 206
Tumminello, Michele, 39
2008 financial crisis, 9, 64, 67,
    232

Ulam, Stanislaw, 5, 6, 85, 208

van Leeuwenhoek, Antonie,
    100
Veblen, Thorstein, 141
von Frisch, Karl, 13
von Neumann, John, 5–6, 23,
    207–208

Wealth of Nations, 5, 30–31
West, Geoffrey, 161, 167
Whitman, Walt, 227

Zinner, Ralph, 92
Zipf's law, see scaling
Zipf, George Kingsley, 165

JOHN H. MILLER is a professor of economics and social science at Carnegie Mellon University's Department of Social and Decision Sciences and an external faculty member of the Santa Fe Institute. He lives in Pittsburgh, Pennsylvania.